D1357851

The New Astronomy

The New

Paul and Lesley Murdin

Astronomy

*black holes, white dwarfs, pulsars,
quasars and supernovae —
how the new astronomy is changing
our concepts of the universe*

Thomas Y. Crowell Company
Established 1834
10 East 53rd Street, New York 10022

The Crab Nebula—arguably the most important object in the sky to the study of astrophysics. Red filaments of hydrogen gas still glow more than 900 years after Chinese astronomers saw a brilliant "guest star" appear in this spot.

Library of Congress Cataloging in Publication Data

Murdin, Paul.
The new astronomy.

Bibliography: p. 211
Includes index.
1. Astronomy. I. Murdin, Lesley, joint author. II. Title.
QB44.2.M87   523   77-5788

Published in the USA 1978 by Thomas Y. Crowell
Company Inc, New York, N.Y.

ISBN 0 690 01474 0

Printed in the United States of America

To our parents

That tho' a man were admitted into heaven to view the wonderful fabrick of the world, and the beauty of the stars, yet what would otherwise be rapture and extasie, would be but a melancholy amazement if he had not a friend to communicate it to.

Archytas (attr.)

Contents

List of tables

I   *Supernovae in space and time*

From time to time bright new stars called supernovae flare briefly in the sky. In fact, there have been only five supernovae seen by man's unaided eye in the last 1000 years. But supernovae have an importance in astronomy which transcends their mere numbers. Why?

The Universe is sending us at this moment virtually all the information that it ever will send us. Anything that we can find out about the Universe we can probably find out now. But understanding does not come easily. The Universe does not arrange itself to make itself comprehensible. All the pieces of all the cosmic jigsaws are there, but scrambled and mixed up with scraps which have no significant place in that picture. Supernovae are a large and significant piece of the cosmic jigsaw.

But even fundamental clues to the structure of the Universe can be overlooked, unrecognized when seen by an uncomprehending eye. For example the appearance for many months in AD 1006 of a supernova which shone so brightly that it cast shadows on the ground apparently did not change the beliefs of those who thought that the heavens were immutable. However two similar bright supernovae in 1572 and 1604 threw light into the corners of minds ready to understand that the stars were not permanent, and the new astronomy began.

At first astronomers concentrated on determining only the motions of the stars and gave fleeting attention to determining their life cycle. This was because although we can talk of stars having birth, life and death, they go through these stages on a very long time scale and changes are not usually apparent. The Sun has been as it is for some four billion years. Over a man's lifetime, indeed, over the lifetime of Man, most stars *have* remained very much the same. But recognizing from the occasional appearance of so-called new stars that the heavens do change, astronomers have sought to understand how stars evolve. It has not been easy.

Astronomers see very many different kinds of stars in the sky; their objective in looking at these stars is first to explain how each kind works and what gives it the appearance that it has to us, and second to see how it changes from one stage to another.

An astronomer has been compared with a Martian presented with a snapshot of a forest. The Martian must try to understand from his photograph the life cycle of a tree. He needs to know how an acorn turns into an oak sapling, and how an oak sapling turns into a fully mature tree; how the tree dies, falls and decays on the forest floor. Without actually seeing the processes of change in a tree, the Martian must study the different organisms that he finds and try to classify them into their different categories, guessing how one changes to another.

In the same way, the astronomer studies the various types of stars, tries to classify them under different headings, and tries to explain how one type of star changes to another. Remarkably, after only 100 years of studying the intrinsic nature of stars, astronomers now believe that they do have a workable system for explaining how one kind of star changes into another. They believe that within a general outline, they have explained the whole lifetime of typical stars: from how they are born and begin to shine, up to the moment when all their energy is used up and they cease shining brightly and die.

It is only very recently that astronomers have begun to study the death of stars. Unlike the majority of a star's life, which is long and peaceful, the stage which marks the death of a star can be brief and dramatic—so brief that it may last just a few months and so dramatic that for a few days the star outshines all its millions of neighbors put together. The lifetime of stars is so long and this death so brief that on average just one such event in a galaxy of 100 billion stars can be witnessed in the lifetime of a man.

The bright explosion which marks the death of a star is a supernova. Where no star was seen before, astronomers see a bright new star shining. When first recognized such objects were named *novae* (pronounced "novee") meaning "new stars," although astronomers now understand that the star is not new but was so faint that it was not noticed before.

In 1935 it was formally recognized that there were at least two very distinct kinds of nova, one much brighter than the other. The fainter kind is probably caused by a relatively weak explosion on one of a pair of stars orbiting each other, just as the Moon orbits the Earth. The brighter kind (which is 100,000 times brighter) is a supernova, and it is this explosion which marks a stellar death.

In our own Galaxy of 100 billion stars only five supernovae have been witnessed in the last 1000 years, none since the invention of the telescope in 1608. But astronomers have found in the sky the remains not only of these supernovae but of others which occurred many thousands of years ago. A supernova explosion produces two visible kinds of objects. At the site where a supernova occurred astronomers see the shell of the exploding star speeding into space in fragments, colliding with tenuous gas in space and glowing from the force of the collision. This is called a supernova remnant. At the center of a supernova remnant may sit the hard core of the star that died, a star so faint that most emit no discerned light, a star packed so tightly by the force of the explosion that a matchbox full weighs a billion tons. It is called a neutron star.

The most-studied example of a supernova and its remnant is known as the Crab. Seen as a bright star in AD 1054, the Crab supernova produced a nebula which was discovered in the 18th century. At the center of the Crab Nebula lies a faint star, the neutron star produced

3

by the supernova. The star is spinning on its axis at a rate of 30 revolutions per second, just as the Earth rotates once per day. A "hot spot" on the neutron star shines like a lighthouse into space and, as the beam passes across Earth once each revolution of the neutron star, it is perceived to flash or pulse. It is a pulsar.

The study of supernovae has shed light on other unexplained problems in astronomy, such as how the elements came to be formed, including those in our bodies, and on the origin of cosmic rays, which are speeding particles of matter in space.

Astronomers believe that supernovae are at the origin of cosmic rays and thus at the origin of part of the natural level of radioactivity on Earth. Some even speculate that past supernovae have played a part, through increasing radioactivity due to the cosmic rays, in the evolution of life itself.

Thus supernovae, worth studying in their own right, have wide-ranging links with other studies as well as occupying a central position in the science of astronomy. Astronomers find it worthwhile to spend time studying not only supernovae in other galaxies, and the remnants of supernovae which have occurred in the Galaxy, but to delve into the tantalizing historical records of past galactic supernovae, attempting to discover the galactic supernovae which caused the remnants they see.

Research into supernovae is only partly a matter of library study. Most information comes from investigating the sky. Practically everything has been achieved that can be achieved just by looking, and since the Universe is sending us all the information it ever will, it is only by new techniques that astronomers can achieve any new understanding. Some of the advances come from building bigger and better telescopes to perceive the light from fainter stars with finer detail. But the optical astronomer looks at the Universe through a restricted window in the atmosphere of the Earth. He does not see cosmic ultraviolet light, since it is absorbed by ozone in the atmosphere, and he does not easily see cosmic infrared light since this is absorbed by water vapor and oxygen. Just before World War II a second window on the Universe was opened: the window penetrated by radio telescopes. The brightest "star" seen by radio astronomers turned out to be a supernova remnant formed just 300 years ago by an unseen supernova. Among the other bright radio "stars" is Taurus A, the radio astronomer's name for the Crab Nebula, remnant of the supernova of 1054. Among the fainter radio stars were discovered the pulsars, now known to be neutron stars formed in supernova explosions.

No further wide windows onto the Universe are available to the Earth-bound observer. To see the Universe of stars through other windows the astronomer flies his telescopes above the atmosphere. The first X-ray "star" to be identified, seen by a rocket-borne X-ray telescope, was the Crab Nebula. But it has been the previously unknown X-ray stars which have been the greatest surprise: the so-called compact X-ray stars are neutron stars, such as the one in the Crab, and more bizarre, the black holes, also formed in supernova explosions.

The contemporary study of astronomy has been described by Geoffrey Burbidge as being divided into the study of the Crab Nebula and the study of everything else. An exaggeration of course, but this remark spotlights the central place held by the Crab and other supernovae in the new astronomy. This book begins with the search for the historical supernovae, and especially the progenitor of the Crab Nebula, the supernova of 1054.

II *Guest stars: the historical supernovae*

The guest star of 1054

In the first year of the period Chih-ho, the fifth moon, the day chi-ch'ou, a guest star appeared approximately several inches southeast of T'ien-kuan. After more than a year it gradually became invisible.

In these straightforward words Toktaga and Ouyang Hsuan, the 14th century Chinese authors of the official history of the Sung dynasty, the *Sung Shih*, noted the appearance of a previously unknown bright star in the constellation now known as Taurus, the Bull. The day referred to is what we would now call July 4, 1054, and the star T'ien-kuan is what present-day astronomers call Zeta Tauri. These prosaic details pin down the precise day on which occurred an astronomical event whose effects are still with us over 900 years later.

To the Chinese, guest stars were well worth noting, and indeed looking out for. They believed that man lived on Earth in a kingdom roofed with stars, and that man's destiny was subject to a "cosmic wind." Chinese emperors appointed court astrologers who watched the sky to ascertain the direction in which this cosmic wind would blow their subjects. These astrologers had been noting down celestial events since the 14th century BC, as these were believed to mark events of great significance in earthly affairs such as the death of princes.

6

The observations were not noted down as incidental asides but as part of a deliberate policy. According to the Jesuit Lecompte's account in AD 1696 of the Ch'ing astronomical bureau: "They still continue their observations. Five mathematicians spend every night on the tower watching what passes overhead. One gazes towards the zenith, another to the east, a third to the west, the fourth turns his eyes southwards and a fifth northwards, that nothing of what happens in the four corners of the world may escape their diligent observation. They take notice of the winds, the rain, the air, of unusual phenomena such as eclipses, the conjunction or opposition of planets, fires, meteors and all that may be useful. This they keep a strict account of, which they bring in every morning to the Surveyor of Mathematics, to be registered in his office."

The medieval Chinese historian Chang Te-hsiang writes in the *Sung hui-yau* that soon after the guest star of 1054 became visible, the Director of the Astronomical Bureau, Yang Wei-te, presented himself prostrate and kow-towing before his Emperor to tell him of its appearance. Perhaps Yang was fearful of not having foretold the coming of the

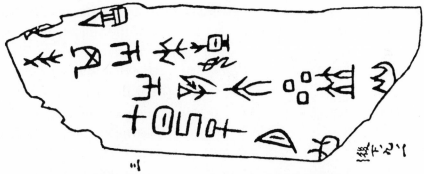

Chinese oracle bones were made from an animal's shoulder blade and inscribed with a question. After searing the bone with a red hot poker, the answer to the question was divined from the pattern of cracks which appeared. The appearance of a guest star might have been held to confirm the answer. This bone dates from 1300 BC and reads: "On the 7th day of the month a great new star appeared in company with Antares." It is the first record of a "guest star" which could have been a supernova.

7

guest star, since he assured the Emperor that because the star did not conflict with the constellation Pi (the nearby Hyades star cluster) and was bright and lustrous, it meant that a person of great wisdom and virtue was to be found in that part of China. This oblique compliment was no doubt well received by the Emperor and his assembled court,

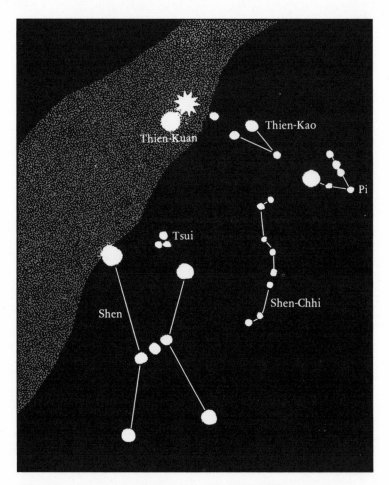

The supernova of 1054 occurred in the Milky Way near the Chinese constellation Thien-Kuan, north of Shen, the present constellation Orion. It " did not conflict with Pi " (the Hyades) noted one astrologer, reassuringly.

8

which acclaimed Yang's remarkably accurate prognostication, for Yang requested that it be filed at the Bureau of Historiography, perhaps the better to be retrieved if he should displease the Emperor at a later date and have to re-convince him of his loyalty.

The account of this episode goes on:

During the third month in the first year of the Chia-yu reign period [March 19 to April 17, 1056] the Director of the Astronomical Bureau reported "The guest star has become invisible, which is an omen of a guest's departure." Originally, during the fifth month in the first year of the Chih-ho reign period, the guest star appeared in the morning in the east, guarding T'ien-kuan. It was visible in the day, like Venus, with pointed rays in every direction. The color was reddish-white. It was seen like that for twenty-three days altogether.

The guest star was also seen and recorded by Japanese astronomers. They noted its appearance, as bright as the planet Jupiter, in early June 1054. It therefore seems that they may have observed it before it was at its brightest (as bright as the planet Venus) which occurred early in July 1054 when it was discovered by Yang Wei-te. Possibly this is why he seemed somewhat apprehensive when making his report to the Emperor. He may well have had a real danger to fear—according to legend, the Chinese astronomers Ho and Hi were beheaded for failing to predict the solar eclipse of 2137 BC.

The star in an Arizona cave

A rather strange thing about the new star of AD 1054 is that no written records of it can be found in the writings of other cultures. European, Arabic and Korean chronicles remain silent about an event which must surely have been an astronomical spectacle, far outshining such pale rivals as Halley's Comet, which appeared 12 years after.

But in North America, despite the lack of writing at that time, there is a strong possibility that the new star actually was seen and recorded. The evidence was uncovered in the early 1950s by two astronomers on a non-astronomical expedition to rescue cultural relics in northern Arizona before the completion of a dam flooded a valley of the Colorado River.

The astronomers were Helmut Abt, then at Yerkes Observatory, Wisconsin, and Bill Miller, who was chief photographer at Mt. Palomar Observatory, California. On their expedition they found two remarkable prehistoric drawings. Later on, similar drawings were discovered at other sites.

The first drawing, in the White Mesa, was found on the wall of a cave and shows a crescent Moon with a circle, apparently representing a bright star, overlapping the lower cusp or horn. In the second drawing the Moon is also shown as a crescent (but reversed from the first) and the circular object is directly under the lower cusp. This drawing was found on a canyon wall near ruins in Navajo Canyon. A third shows a series of circles below the crescent possibly representing motion of the Moon with respect to the bright star.

Because the first drawings were found in association with Pueblo Indian dwellings it is possible to give some indication of their date by archaeological methods. Pueblo Indian pottery from different locations and of different dates is highly distinctive. Potsherds collected at the two sites have been analyzed by Robert C. Euler, Curator of Anthropology at the Museum of Northern Arizona, who found that most of the inhabitants of the White Mesa site lived there later than AD 1070, but that there were earlier sherds in the collection. At the Navajo Canyon site a deep arroyo, or gulley, cut into the floor of the canyon, had exposed broken pieces of pot, with the oldest at the deeper levels. Euler determined that the site was occupied by Pueblo Indians from before AD 700 to after AD 1300 with about a fifth of the sherds collected dating from between AD 900 and 1100.

It is possible that the drawings represent nothing rarer than a near approach of the crescent Moon to one of the brighter planets, Venus or

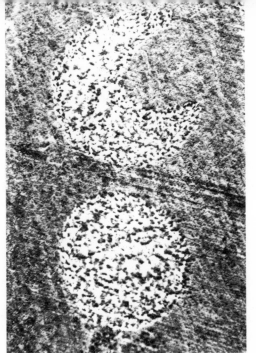

Contemporary views of the 1054 supernova? Two ancient Indian markings discovered by William C. Miller. He found the one at left at the White Mesa of Arizona in 1952 and the one at the right at Navajo Canyon in 1954.

Jupiter perhaps. Such a so-called conjunction between the Moon and a bright planet takes place several times a year; it is an event which has for a long time attracted the attention of astrologers from many cultures because the Moon and planets are supposed to exert influences which augment or conflict at times of conjunction. The crescent Moon and stars is therefore a common symbol; there are more than a dozen representations on present day national flags—Turkey's, for example.

There would have been many times when such a conjunction occurred during the centuries in which the Arizona sites were occupied. If this recurrent event was of interest, however, it is strange that just a few pictographs have been found in Pueblo Indian sites. On the other hand, the scale of the circle relative to the Moon in both drawings suggests that the star was comparable in brightness to the crescent Moon, and certainly the guest star of 1054 was one of the most notable stars known to have appeared during the time the sites were occupied. Cambridge astronomer Fred Hoyle suggested this to Bill Miller, who then calculated the position of the Moon during July and August 1054. He found that on July 5, 1054 at 3 o'clock in the morning (that is, the day after the guest star was noted at its brightest by Chinese astronomers) the crescent Moon would have appeared just above the guest star in the morning twilight, and could have been seen from the Pueblo Indian homes, which had an unobstructed view to the east.

The evidence that the drawings are representations of the new star of 1054 is circumstantial and there are some difficulties—why, for instance, is the crescent Moon reversed in one picture? (Miller suggests that non-astronomically trained people do not worry about this kind of thing and remarks how common it is to find the Moon shown in the wrong orientation in modern illustrations.) But it may well be that these drawings really were made by Pueblo Indians 438 years before Columbus sighted land in the Americas, and are indeed records of what amounts to a remarkable pre-Columbian Independence Night fireworks display.

What can be learned about the nature of the guest star from the ancient observations? We have first of all its approximate position near to the star Zeta Tauri.* But more than this, the *Sung Shih*, quoted at the beginning of the chapter, says in another passage about the guest star that it "remained" in the sky near Zeta Tauri for almost two years.

The ancients recognized two kinds of stars, the fixed stars and the wandering stars, or planets. It is now known that in fact both the fixed stars and the planets are moving through space at speeds comparable with one another. But the wandering stars—planets—appear to move faster because they are much closer to us than the so-called fixed stars. The planets, in fact, belong to the nearby solar system, whereas even the closest star is 10,000 times more distant than the farthest planet.

Similarly, comets too speed through the solar system and wheel across the sky, typically in a few months. So even though the Chinese

---

* The present system of naming bright stars uses the Latin constellation name and a Greek letter, with Alpha (α) usually denoting the brightest, Beta (β) the second brightest and so on. Zeta Tauri is therefore the sixth brightest star in the constellation of Taurus, the Bull.

called some comets "guest stars" and did not exclusively use the term for new fixed stars, the fact that the guest star was noted to be stationary for two years places it certainly as far as the edges of the solar system, beyond the most distant planet, Pluto, and probably outside the solar system altogether.

The Chinese and Japanese records also tell us something of the change in brightness of the guest star. We can deduce from the extracts quoted above that a month before the maximum brilliance of the guest star it was as bright as Jupiter; that at its brightest it rivaled Venus, visible during daylight; that the guest star ceased to be visible in daylight 23 days after that; and that it was finally fading from the evening sky 630 days later.

Because the guest star faded from view to the naked eye into the evening twilight in 1056, it probably was not as faint as the faintest stars visible on the darkest nights, but somewhat brighter.

The evidence that we have of the length of time the star was seen is all consistent with describing this guest star as a supernova, rather than as a nova, since novae fade more quickly.

It is now known as the supernova of 1054. It was one of only five supernovae which can have been seen by the unaided eye of man within the last thousand years.

Just as the guest star faded from view, so its memory faded from men's minds. But the vital details of its visibility had been recorded, were dutifully stored and recopied by Chinese scribes over the centuries, and can now help in the attempt to understand the mysteries of the Crab Nebula, left over from the original momentous explosion.

Sweeping, bushy stars eliminated

What other supernovae were recorded among the oriental observations? To distinguish records of supernovae from records of other celestial phenomena such as comets, aurorae, meteors, lightning and so on, astronomer David H. Clark and historian F. Richard

13

Stephenson have united in an interdisciplinary study. They have isolated all mentions of guest stars which had no motion and so excluded
moving objects such as comets and meteors. Many objects could be
eliminated as being described as "bushy stars" or like a broom ("sweeping stars"), references to a comet's tail (though the records are not
always unambiguous on this point). This gave them a list of 75 probable
novae or supernovae sighted between 532 BC and AD 1604. Among these
they identify six supernovae which occurred before AD 1500 (Table I).

TABLE I   *The historical supernovae*

Year (AD)	Constellation	Duration	Maximum Magnitude	Remnant
185	Centaurus	20 months	−6?	MSH 14 −63
386	Sagittarius	3 months		GII.2 −0.3?
393	Scorpio	8 months		G348.7 +0.3?
1006	Lupus	Several years	−9	P 1459 −41
1054	Taurus	24 months	−5	Crab
1181	Cassiopeia	6 months	+1?	3C 58

Clark and Stephenson recognized them because they are noted in the
records as lasting longer than an ordinary nova and also were seen along
the central line of the Milky Way, which is where supernova remnants
are found and hence the only place where supernovae occur. Apart from
the guest star of 1054 there are two supernovae, recorded in oriental
observations, whose history is known in sufficient detail to have enabled
radio astronomers to claim with some confidence that they have identified their remnants. They are the supernovae of 1006 and 1181.

The supernova of 1006

Probably the brightest star to have appeared in the sky for
the last thousand years was the supernova of 1006 which blazed from

the southern constellation of Lupus. Because it was so far south and was below the horizon to northern European observers, our main sources of information are Arabic, Japanese and Chinese texts. These texts agree fairly well on the position of the phenomenon but it can readily be admitted that they are obscure over other details of its appearance. In spite of the difficulties in interpreting and collating the ancient sources, modern astronomers are convinced that the star was a supernova.

The main eye witness source is the Egyptian Ali b. Ridwan (who died in 1061), who lived in the old city of Cairo. He mentions the supernova in an autobiographical footnote to an astrological work by Ptolemy, the *Tetrabiblos*, which he was editing. He says that it appeared "at the beginning of my education" in the 15th degree of Scorpio. The word he uses for the star is *nayzak* which is fairly rare in Arabian astronomical works. Where it does occur, it refers to a very bright comet, which is how later commentators translated the word. Ali, however, says that the object remained stationary relative to the other stars, while the sun moved into Virgo. Therefore, it was not a comet.

Ali included a list of the exact positions of the planets at the time of his first sighting the supernova. From his list we can be quite sure of the date: April 30, 1006. The modern computation of the planetary positions on this day agrees well with Ali's own data. Ali also includes the expected astrological conclusions. In this case, famine, death and pestilence broke out. Ali detailed his observation of the supernova:

It was a large nayzak, *round in shape and its size two and a half or three times the size of Venus. Its light illuminated the horizon and it twinkled a great deal. It was a little more than a quarter of the brightness of the Moon.*

It is not clear what Ali means when he says that the supernova was about three times the "size" of Venus. Astronomers then thought that the brighter stars had perceptible disks, this being a physiological effect in the eye. Two Chinese sources even say that it was a half Moon.

Ibn al-Jawzi, another Arab source but from the 13th century, says that "it was a large star similar to Venus." It seems probable that these astronomers, in comparing the supernova to Venus, were trying to give an impression of its brightness compared with the brightest star they knew. Ali, in fact goes further and says it was a quarter of the brightness of the Moon. This may mean a quarter of the brightness of the Moon when only a quarter is illuminated. This latter interpretation is more consistent with the comparison to the size of Venus.

Ali and Ibn offer the seemingly independent observations that the light from the star "illuminated the horizon" and that "its rays on the earth were like the rays of the Moon." A Chinese observer noted in the *Sung Shih* that it "cast shadows." Another Chinese source says that "it shone so brightly that objects could be seen by its light." This all indicates that it was much brighter than Venus.

The most significant European account is from Hepidannus, a monk of St. Gall in Switzerland. In his Latin Chronicle of 1006 he says that he saw the star in the extreme south. This places a limit on how far south it could have been since it must have been above the horizon as seen from St. Gall (latitude $47\frac{1}{2}$ degrees). Chinese and Japanese observers place it on the present day border between the constellations Lupus and Centaurus. Its position can thus be tied down to within fine limits.

The color is doubtful. One Chinese observer, the Director of the Astronomical Bureau at the Imperial Court of the Emperor Chen-tung, called it "a large star, yellow in color." This is not to be relied on, however, as an objective description. The reason is partly astrological and partly political. When the supernova appeared, it was so striking that everyone in the Chinese capital, Kaifong, was filled with alarm, and the general opinion was that it was a very bad omen which would be followed by famine and plague. The Director of the Bureau of Astronomy, Chou K'o-ming, was out of town at the time. When he returned, he found the Emperor very anxious and distressed by the situation. Having considered the evidence, he announced that the star belonged to the astronomical category *Chou-po*. This was an excellent move as such a star was an omen of prosperity and could occur only in the reign of a wise and just monarch. The Director was later promoted to Librarian and Escort of the Crown Prince. The important characteristic of a *Chou-po* star from our point of view, however, is that it was always yellow in color. The description is therefore inevitable, given the classification, and does not provide us with scientific information. A Japanese report that it seemed to be blue-white may be more objective.

When we ask how long the supernova was visible, we find difficulties, though it was long-lasting. The Chinese *Chu-Su* says that it "later increased in brightness" but does not say later than what. Ali, however, says that it disappeared suddenly. We are told by several writers that the supernova was visible for three and a half months after which it was too close to the Sun to be seen, but that would cause a gradual disappearance, so probably Ali is not referring to this. Venus in the same position would be clearly visible in daylight; therefore we can say that the supernova at that point could not have been brighter than Venus.

After seven months behind the Sun's glare, the supernova reappeared in the dawn sky between November 24 and December 22. How long after this the star remained visible is difficult to determine, but it seems to have been more than a year. The Chinese chronicle, the *Sung Shih* refers to "a *Chou-po* star" in November 1006 and again on May 15, 1016. No positions are given but if both stars are the same one it must have been erratically visible for up to 10 years. The only indication that both the references are to the 1006 supernova is that these are the only instances of "*Chou-po*" being used to describe a guest star. The *Sung Shih* gives more evidence that the supernova was visible for several years. In a passage specifically describing the supernova of May 1006, it mentions the first disappearance and reappearance and continues with the significant word "thereafter" to say that it disappeared near the Sun in the eighth month and reappeared in the eleventh month. Because this suggests a continuing phenomenon, being hidden by the Sun annually, the star must have been visible for at least two years.

We may be trying to give too precise a description from those fragmented notes recopied by generations of scribes and summarized by medieval historians from the bits and pieces salvaged after the Mongol invasion of China in 1345. Nevertheless, the location is relatively clear and has enabled two radio astronomers, Frank Gardner and Doug Milne, to identify in 1965 a radio source at the area where the 1006 supernova appeared. Using the 210 foot radio telescope at Parkes in New South Wales, Australia, they found a structure closely resembling other supernova remnants.

In 1957 Walter Baade had tried to find visible traces of gas where the supernova had appeared, just as he had previously found the nebulae left by other supernovae, but was not able to make any positive identification. The object was too far south for the Californian telescopes. In 1976, however, Sidney Van den Bergh, using the Cerro Tololo telescope in Chile where the constellation of Lupus passes overhead, discovered a faint wispy nebula close to the radio source, ejected from the supernova when Ali b. Ridwan was a student.

The supernova of 1181

Another supernova flared up and briefly amazed observers in AD 1181. The evidence is found in Chinese and Japanese chronicles and is difficult to interpret. Finding out as much as possible is worth some effort, as radio astronomers have found a supernova remnant called 3C 58 in a position which the records indicate was the position of the 1181 supernova. We must fit together, as well as we can, the observations of the ancient astrologers and astronomers with the work of the 20th century radio astronomers, in order to form a picture of the violent death of the star and its subsequent gradual dissipation into the interstellar matter.

The records present us with the two most common difficulties of this sort of work. One is that they do not agree. The other is that what they do say is imprecise. The star was observed from three geographical areas and we can make some deductions from the three sets of records.

18

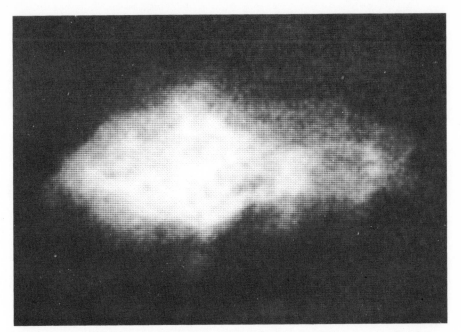

This radiophotograph of 3C 58, the remnant of the supernova of AD 1181, has been made from observations with the Westerbork Synthesis Radio Telescope by A. S. Wilson and K. W. Weiler. Like the Crab Nebula it has a flat appearance, rather than the shell-like look of most other supernova remnants. Four times more distant than the Crab, it is almost six times its size although of comparable age, and so has exploded at six times the average speed of the Crab. No optical nebula has been found at this site.

The first to see it were the observers in southern China. The *Sung Shih* tells us that the star was first seen on August 6, 1181. The next to see it seem to have been the Japanese. The *History of Great Japan,* written in 1715, says that a guest star appeared in the north on August 7, 1181. The Chinese in the northern Chin empire reported in the *History of the Chin Dynasty* seeing the star on August 11.

How long the star was visible is of critical importance in deciding whether or not it could have been a supernova identifiable with the known radio source. The *Sung Shih* says that the star was visible until February 6, 1182 "altogether 185 days; only then was it extinguished." The account in the *Chin Shih* gives a somewhat shorter time, 156 days. The duration of several months makes it a reasonable assumption that the star was a supernova.

The three groups of accounts state that the supernova appeared in or near different Chinese constellations. However, all lie within the present-day constellation of Cassiopeia. Chinese constellations were not firmly defined and the discrepancies are not significant, just disappointingly vague.

The evidence for the color and brightness of this star is largely Japanese. The *Azuma Kagami* tells us:

At the hour hsu [*19–21 hours local time*] *a guest star was seen in the northeast. It was like Saturn and its color was bluish-red and it had rays. There had been no other example since the third year of Kanko* [*the supernova of* AD *1006*]

The comparison with Saturn is striking because Saturn would not actually have been visible between those hours, but only towards dawn. Mars would have been visible at the time, and therefore the mention of Saturn seems to deliberately imply that the supernova's brightness was closely comparable. We do not know, however, whether this was the supernova's maximum brightness or whether it was discovered before maximum. Furthermore, this Japanese chronicle says that there had been no other star of its kind since 1006. The supernova of 1006, as we know, was very bright. Perhaps the supernova of 1181 became very bright too. This interpretation would be consistent with the considerable interest which the star aroused and the number of references to it. But the evidence is not clear enough, and presumably never will be.

The Star of Bethlehem

The best known bright star in history is the Star of Bethlehem. Was it a supernova? To find out, we have to examine the documentary evidence in the same way as we have done for the Chinese and Arabic accounts of other supernovae.

At the birth of Jesus, according to the gospel of St. Matthew, chapter 2, "there came wise men from the east to Jerusalem saying, where is he that is born King of the Jews? For we have seen his star in the east and are come to worship him." The term translated here as "in the east" means more precisely "at its heliacal rising," that is, the wise men saw the star appear in the first rays of dawn.

From Jerusalem the Magi, who may have been astrologers from Persia or from the Tigris-Euphrates valley civilizations of Assyria, Mesopotamia, or Babylonia, traveled south to Bethlehem following the star which "went before them, till it came and stood over where the young child was." This passage is difficult to identify with any astronomical phenomenon since the motion of stars is generally east to west and astronomical objects are so distant that they do not identify one particular terrestrial location but stand equally over whole areas when at the zenith.

Setting this aside for a moment, however, the third item of evidence is from the Protoevangelium of James (21:2), one of the Apocryphal gospels not included in the Bible, which offers the following:

And he [Herod] questioned the wise men and said to them: "What sign did you see concerning the new-born King?" And the wise men said: "We saw how an indescribably great star shone among these stars and dimmed them, so they no longer shone, and so we knew that a King was born for Israel."

Though specific as to the brightness of the star, which would be comparable to the brightness of the Full Moon if it flooded the sky with its light and rendered surrounding stars invisible, this passage raises the difficulty as to how Herod and his advisers could have come to miss

21

noticing the star, unless Herod's question is deliberately disingenuous.

When did the birth of Jesus occur? The first naive attempt at an answer is December 25 in the first year of the Christian era. However, the tradition that Jesus was born at midwinter began about AD 336, possibly because Christians wished to hide their celebrations among the general festival of Saturnalia, or, more likely, because the Church drew the existing pagan festival within the Christian tradition. Luke (2:8) says that at the time of Jesus' birth, shepherds were "abiding in the fields, keeping watch over their flock by night." During winter in Judea flocks were penned, being set free in the spring and guarded by night in the lambing season (March and April).

As to the year, the presently accepted calendar of years is from a correlation between Christian tradition and Roman imperial history which is expressed in a calendar reckoned *ab urbe condita* (AUC), from the founding of the city of Rome. The correlation which has been adopted is by Dionysius Exiguus (AD 525) who missed the year zero between 1 BC and AD 1 and forgot the four years during which Emperor Augustus ruled under his own name of Octavian. This puts Jesus' birth in 5 BC or AUC 749. Herod died just before passover in 4 BC. Jesus was therefore certainly born before then. In his account of Jesus' birth, Luke says that Caesar Augustus had ordered a tax and that this was why Jesus' parents had to travel to Bethlehem. Such an order was issued in 8 BC; the tax would have been collected in the years following. Luke says that tax collection was begun when Quirinius was governor of Syria, but he was not governor until AD 6, although he was an Emperor's legate in Syria between 6 and 5 BC, and Luke may have been confused as to his rank. Tertullian, a Roman historian, says that the census at the time of the birth of Jesus was taken by Saturninus, who governed Syria between 9 and 6 BC.

It seems that the birth of Jesus occurred one springtime between 7 and 5 BC. This excludes Halley's comet as the star (it appeared in the autumn of 12 BC). Another significant astronomical event was a conjunction of Saturn and Jupiter in the constellation Pisces in 7 BC. First suggested as the possible Star of Bethlehem by Kepler, this was a similar conjunction to the one which led to the discovery of Kepler's supernova.

The possibility that the Star of Bethlehem was a supernova (or nova) is one which occurred to astronomers including Tycho Brahe after the appearance of the supernova of 1572, which was interpreted by some as signifying a further event of the same kind, possibly the second coming of Christ. What was possibly a nova or supernova was recorded in the *History of the Former Han Dynasty* as occurring in late March or early April in 5 BC, and lasting over 70 days. It appeared in what is now called the constellation of Capricorn. In springtime this constellation rises some five hours before the Sun so that the star would have been first observed in the rays of the dawn, as Matthew implies. The relatively short length of time for which the star was visible suggests that it was a nova rather than a supernova. The Chinese records actually call the object a "broom-star" which is usually used for comets having tails like a brush, although, as pointed out by D. H. Clark, J. Parkinson and F. Stephenson, one record of the well known supernova of 1572 misclassifies this star in the same way. Whether nova or comet, it was probably not a supernova.

The notion that the Star of Bethlehem was a nova or supernova survives in modern literature in a story (*Nova*) by Arthur C. Clarke. In Clarke's story, a Jesuit astronomer/astronaut finds his faith severely shaken by the discovery of the archaeological remains of a beautiful, peaceful and cultured civilization, exterminated when their sun exploded in a supernova explosion. His astronomical training leads him to calculate the date of the catastrophe, only to find that it was apparently timed by God to occur so that the light of the explosion reached Earth at just the right moment to proclaim the birth of His Son.

Apart from fiction, contemporary chronicles of the Chinese or other peoples might have been expected to record the supernova, if there was one, especially if it was as bright as the Full Moon. No such records have been found. Furthermore, a supernova would not have had the unusual motion attributed to it in Matthew 2:9.

23

It is of course possible that the Star of Bethlehem was a genuinely miraculous event for which there is no physical explanation, in which case the above analysis is in vain. It is also possible that the narrative is what in Jewish tradition is called a *midrash*, a historical equivalent of the Christian *sermonizing*, that is, a historical fact presented in a popular manner with decorations adapted to the reader's expected mentality, with echoes of previous parallel events such as, in this case, the birth of Moses or Abraham. British astronomer David Hughes writes "no king worth his salt in those days was born without some celestial manifestation. A star greeted the birth of Mithridates (131–63 BC) and Alexander Severus." This is probably the way to reconcile the difficulties in the various records, which do not have the physical self-consistency of, say, the Chinese accounts of the supernova of 1054, and cannot be taken to compel the same conclusion.

The Star of Bethlehem might perhaps have been the nova of 5 BC but there is no evidence that it was a supernova.

Since bright supernovae do not often explode within sight of the Earth, and a man is unlikely to see one in his life, the impact that they made on the ancient world may perhaps have been much greater than surviving records imply. In the Renaissance, however, two circumstances combined to make the modern astronomer look there with particular interest when on the track of supernovae. One was the appearance very close together of two spectacular supernovae. The second was a new interest in observing and recording natural phenomena for their own sake. For these reasons, the Renaissance supernovae were of immense significance both at the time and to us now.

III *The Renaissance supernovae: shattering the crystal spheres*

Despite the evidence of their own eyes, most people today believe what astronomers tell them about the way the world moves. From an early age we are all taught that the Earth is round, not flat, and that like the other planets it spins on its own axis and orbits the Sun. Yet every day we see the Sun move through the sky, along with the stars and planets. Few have time to prove to themselves that the Earth really does move round the Sun.

Four hundred years ago, astronomers assured those who listened that the Earth was stationary, and orbited by the Sun. Our ancestors just 20 generations away believed this. Supernovae helped change their minds.

Crystal spheres and epicycles

For 2000 years, most Europeans accepted the description of the Universe that had been given by the Greek philosopher Aristotle (381–322 BC). Ptolemy of Alexandria turned Aristotle's notion into a quite workable mathematical form in the 2nd century AD. Then in the 13th century, Thomas Aquinas took Ptolemy's system as the correct picture of the Universe to give a background to his ideas on theology.

When Aquinas' theology was accepted by the Roman Catholic

Peter Apian's picture of the Universe, published in 1524, illustrates the Aristotelian conception of the Universe fused with Christian theology. Outside all is the Coelum Empirreum, *the highest heaven, home of God and the Elect. Ten spheres are nested within, starting with the* Primum Mobile, *inside which are the ninth sphere of crystal, the eighth sphere containing the Firmament of fixed stars and the spheres of the seven planets, including number 4, the Sun. Below the sphere of the Moon are the terrestrial regions of fire, the cloudy sky and, at the center of all, the Earth itself.*

Church, the cosmologies of Aristotle and Ptolemy became woven into the social order which was seen as a chain stretching from God down to the humblest living creature. A king might hope to be secure on his divinely granted throne because order was seen in the heavens. The philosophy and the science were not questioned because if one group of certainties was shattered, all the rest, including the social order, would be subject to doubt.

What was this cosmology which so constrained people's lives? In the Ptolemaic system, the Universe has a set of spheres like an onion. The Earth was thought to be at the center of the visible Universe. Completely surrounding the Universe, according to medieval philosophy, was the Empyrean where God sat enthroned, accompanied by the souls of the just. Within this lay the *primum mobile*, the First Mover. This was a sphere of an intellectual substance which was thought to be the cause of all movement in the heavens. Movement in the heavens was, in turn, the cause of all movement on Earth.

Inside the *primum mobile* was the eighth sphere, or firmament, the sphere embedded with the fixed stars. (Some saw the necessity for a ninth sphere, the *caelum igneum*, or fiery sphere, to account for some of the apparent movements of the planets.) Below these outer spheres were the spheres of the seven planets, as Saturn, Jupiter, Mars, the Sun, Venus, Mercury and the Moon were then called. Each revolved at its own speed about the Earth, the spheres closest to Earth slipping behind the rotation of the firmament by the largest amount.

Aristotle said that the sphere was the perfect geometrical form. This was why the heavenly bodies were embedded in spheres and why they logically had to move in circles. The lowest sphere contained the Moon and marked the boundary between celestial perfection and the sublunar region of death and corruption which mankind inhabited. With this picture, everything falls into place. Thomas Aquinas was able to link it with the Christian view of the Fall of Man and his propensity to sin.

Even though the English philosopher Thomas Digges believed, like Copernicus, that the Sun was the center of the Universe, he still held to

27

the theory that above the Moon all was unchangeable and that death and decay were possible only beneath, describing in 1576

the Moon's Orb that environeth and containeth this dark star [Earth] and other mortal, changeable, corruptible Elements.

John Donne (1571–1631) often referred to these heavenly spheres, even in love poetry:

> *If, as in water stir'd more circles bee*
> *Produc'd by one, love such additions take,*
> *Those like so many spheares, but one heaven make,*
> *For they are all concentrique unto thee.*

But at the center of the spheres, wrote Donne, were "dull sublunary lovers" whose love could not survive absence because it was based on the senses alone and was therefore imperfect, in contrast to God's perfect love from heaven, above the highest celestial sphere.

Given the preconception that above the Moon everything was perfect and eternal, the followers of Aristotle accounted for any changes which they saw occurring in the sky by insisting that they took place below the Moon. Comets, for example, were said to be atmospheric. According to Aristotle himself, comets were exhalations from the Earth produced by the burning of gases in the atmosphere above the Earth and set on fire by the Sun. Today's view, that they are celestial bodies traveling on elongated orbits which bring them in from the depths of space through the solar system, between the planets, would have been impossible in the Ptolemaic system. The comets would have had to penetrate the crystal spheres in which the planets were embedded, shattering the perfection of circular motions. In Aristotelian science, comets were in fact classed with rainbows, gales, dew, lightning and all other atmospheric phenomena as being meteorological. It was a much later development which restricted the use of the word *meteor* to shooting stars.

28

Although astronomical historian Owen Gingerich has cast doubt on this picture's historical authenticity, the medieval cosmology of the Universe in which the Earth was at the center of a sphere of stars is well illustrated in this alleged fifteenth century woodcut. A pilgrim looks through the Firmament at the mechanisms of the Primum Mobile beyond.

Curiously, although oriental astronomers recorded dozens of novae, only two possible novae were known to European astronomers before 1572. No systematic way had been developed to explain how such a temporary phenomenon could appear in the eternal firmament.

Each of the two possible novae was explained away. The first was the Star of Bethlehem, but because it was thought to be genuinely miraculous it needed no general explanation. Secondly, the Greek historian Pliny records that Hipparchus is said to have observed a new star in 134 BC. The usual explanation for this was that what he saw was actually a comet, although Hipparchus should have known the difference between the point-like appearance of a star and the diffuse appearance

29

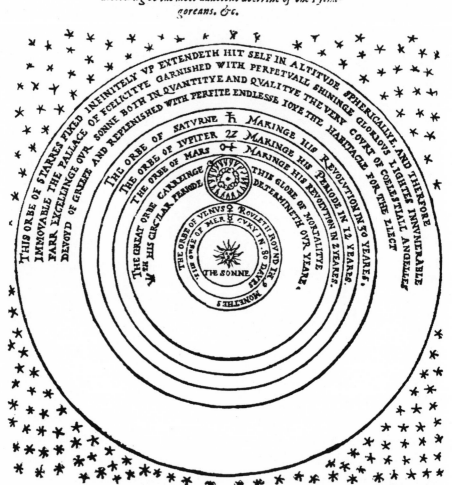

Thomas Digges' Perfect Description of the Celestial Orbs *was published in*
1576. While trying to come to terms with Copernicus' idea that the Sun was at
the center of the Universe, Digges was still a prisoner of the Aristotelian idea that
the stars were immutable. The Orb of stars is described as "immovable . . .
garnished with perpetual shining glorious lights innumerable," whereas the Orb
of the Earth carries "this globe of mortality."

of a comet (the word comet derives from *coma*, meaning hair). However, it was firmly held that whatever he saw, it was sublunary.

So pleasing was Aristotle's description of the Universe, with its perfect circles and uniform speed of planetary motion, all within the constant firmament, that until the 16th century, and even after, most philosophers expended their energy in *saving the appearances*, that is, devising new methods of calculation and minor embellishments of Aristotelian principles which would enable the observations to be reconciled with the theory.

Ptolemy in fact found that the movements of the planets which he observed did not fit the theory of exactly circular orbits. He elaborated Aristotle's theory by claiming that planets moved on *epicycles*, or small circles whose centers themselves moved in circular orbits. In a similar way, Ptolemy dealt with the problem that the observed speeds of the planets were not uniform at all points of their orbits, as according to Aristotle they should have been, since uniform motion was the perfection expected in a cosmic body. Ptolemy maintained that the speed of a planet was indeed constant, not as seen from the Earth, but as observed from another point in space, the *equant*.

Later philosophers followed in Ptolemy's footsteps, devising elaborations of the epicyclic idea to "save the appearances." Even Copernicus who in 1543 took the drastic step of re-ordering the solar system so that the Sun, and not the Earth, was at the center of the Universe, still presented his new theory of the solar system in terms of epicyclic motion.

By about this time it was becoming increasingly clear that the motions of the planets were getting harder to explain on a simple epicyclic system. More and more complexities had to be introduced in order to predict accurately the positions of the planets in the sky. But the ancient learning, supported by the Church, was deeply entrenched. Perhaps it is not surprising that philosophers were prepared to go to great lengths to preserve the old system, which to most people must have seemed the only logical one.

Indeed, when Andrew Osiander came to write the preface to Copernicus' revolutionary book, he had to say that the new theory merely simplified calculations and did not necessarily mean that the Sun had ousted the Earth from the center of the Universe. But nonetheless the contradiction between theory and observation arose because of the Aristotelian presumption that the Universe above the Moon was perfect. It took the supernova which suddenly exploded in 1572 to shatter the crystal spheres.

Observations of the supernova of 1572

The first recorded observation was made on November 6, 1572, by a Sicilian mathematician, Francesco Maurolyco, who observed a very bright, previously unknown star in the constellation of Cassiopeia. There was great excitement about the new star. "I am unable to admire enough the new shining of the star of our time" wrote Maurolyco. He noted that he saw the star at the third hour of the night and wrote down its approximate longitude and its angle above the horizon. There is some doubt about whether Maurolyco had seen the supernova before November 6, but a Spanish philosopher, Hieronymus Mugnoz, was teaching an outdoor class in astronomy on November 2 and said afterwards that he would certainly have noticed the new star in Cassiopeia if it had been visible then. The beginning of the supernova can therefore be placed between November 2 and 6, 1572.

On November 7, Paul Heinzel of Augsburg, Bernhard Lindauer of Winterthur in Switzerland and Michael Mästlin of Tübingen, Kepler's teacher, also saw the supernova. Professor Mästlin satisfied himself that the nova was a star and not a comet. He did this by selecting two pairs of stars in Cassiopeia so that lines between the members of each pair would intersect at the supernova, which was accomplished by holding a thread before his eyes so that it passed through two of the known stars and the new star. In this way, he was able to say that the new star was not moving in relation to the other stars of Cassiopeia. Thomas Digges carried out the same experiment using a six-foot ruler, and placed the

Tycho Brahe

star at the intersection of the line joining Beta Cephei to Gamma Cassiopeiae and Iota Cephei to Delta Cassiopeiae.

Mästlin's and Digges' simple observations were elegant, but the man who won fame and fortune from the supernova of 1572 was Tycho Brahe. Brahe, a Danish astronomer, was an extraordinary man. His personal life was wild and undisciplined in the extreme. He had a gold and silver bridge to his nose necessitated by an injury suffered in a duel when he was a student. Later in his life he lost the private island observatory which had been granted to him by King Frederick because of his arrogant and unjust treatment of his tenants. As a scientist, however, he was entirely different. Far from being arrogant and bombastic, he was meticulous and precise. Indeed, his observations, made before the introduction of the telescope, are recognized as the finest ever made with the naked eye, and achieve an accuracy limited only by the acuity of the eye itself.

Brahe was at the beginning of his career in 1572, and it was in fact the supernova which inspired him to devote his lifetime to making accurate

33

measurements of the positions of the stars and planets. As Kepler, his pupil, said, "if that star did nothing else at least it announced and produced a great astronomer." Brahe's book *De Nova Stella* (1573), in which he first set down his observations and the conclusion that he drew "about the new star," caused immense interest and some horror at what were seen as sensational ideas.

In *De Nova Stella*, Brahe described his first sight of the supernova:

Last year in the month of November, on the 11th day of that month, in the evening, after sunset, when according to my habit, I was contemplating the stars in a clear sky, I noticed that a new and unusual star, surpassing the other stars in brilliancy, was shining almost directly above my head; and since I had, almost from boyhood, known all the stars of the heavens perfectly (there is no great difficulty in attaining that knowledge), it was quite evident to me that there had never before been any star in that place in the sky, even the smallest, to say nothing of a star so conspicuously bright as this.

This cool account is somewhat at odds with Brahe's further admission that, doubtful of the evidence of his eyes, he sought confirmation from his servants and some peasants driving by that they too could see the new star. They could.

The unmoving star

Brahe's most important measurements of the supernova of 1572 were of its position. Although he did not have the advantage of the more accurate instruments which he later acquired for his observatory on the island of Hveen, he did have a large and well-made sextant-type instrument which he had just finished making. He was able to measure the distance of the supernova from the nine principal stars of Cassiopeia, making measurements as accurately as possible with the naked eye. He repeated the measurements at every opportunity, often several times throughout the night. In fact, the star was sufficiently near to the sky's

34

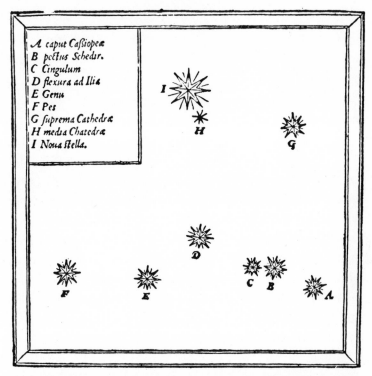

Tycho's map of the 1572 supernova shows the brighter stars of Cassiopeia with the new star, marked "I," brilliantly outshining them. His Latin labels identify the stars by their positions in the mythological "Lady in the Chair" that the constellation is supposed to represent. Thus "A" is the head, "E" the knee and so on.

north pole, the star Polaris, so that from Denmark the star never set and could be kept under observation the year round. Like Mästlin and Digges, Brahe found that its position was unchanged for all this time, from hour to hour, from day to day, from month to month to within the accuracy of his sextant. From other measurements of the positions of stars, we know that Brahe's measurements repeated to an accuracy of a few minutes of arc or less than a tenth of a degree (about the size of a nickel or a British penny held at a distance of 40 yards). Brahe's measure-

35

ments firmly put the supernova unmovingly among the other fixed stars.

What motion might Brahe have expected? It was natural for 16th century astronomers to compare the supernova with other transient phenomena, such as comets. Comets move, characteristically right across the celestial sky in a few months or even weeks. If the new star had been moving at such a rate, Brahe would have detected its motion in a matter of hours, unless, as some of his contemporaries implausibly argued, it was moving directly away from or towards Earth.

He detected no motion in 18 months, eliminating any idea that the new star might have been associated with a planet, since the farthest then known, Saturn, would have moved with motion detectable by Brahe in a week. These general arguments, as well as the observation that the star twinkled just like other fixed stars, in contrast to the planets which shine without twinkling, were used by Brahe to show that the supernova was indeed a star in the eighth sphere, the firmament. But Brahe had specific arguments to show how distant the star was. His measurements with his sextant showed that the star had no *parallax*, or apparent motion caused by the motion of the Earth, and was certainly beyond the sphere of the Moon. Let us see how he was able to prove this. We will give the argument in terms of the Earth's rotation, although Brahe would have assumed the Earth to be stationary, and the firmament to be rotating.

Brahe measured the position of the supernova from Heridsvaad when the star was almost overhead at the zenith in the evening sky (Fig. 1). 12 hours later, as morning approached, the Earth had rotated halfway round and, since the supernova was circumpolar and could be seen all night, Brahe was able to repeat his measurement.

Brahe would be making the second measurement from a new position in space carried there by the rotation of the Earth. If the supernova were close to the Earth, as drawn in Fig. 1, it would no longer be in the same direction. The distance of the supernova is measured in fact by the angle by which it shifts its apparent position, the so-called *parallax*, of the supernova. The smaller the parallax, the more distant the star.

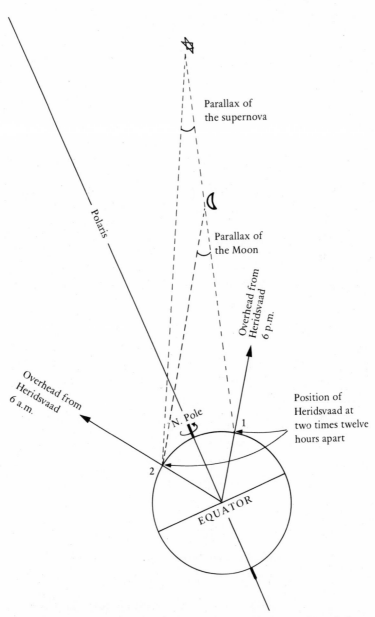

Parallax of
the supernova

Polaris

Parallax of
the Moon

Overhead from
Heridsvaad
6 p.m.

Overhead from
Heridsvaad
6 a.m.

N. Pole

Position of
Heridsvaad at
two times twelve
hours apart

2

1

EQUATOR

Brahe's measurement of the parallax of the supernova showed that it was less than the parallax of the Moon and hence the supernova was more distant.

37

Brahe calculated that at the distance of the Moon the parallax of the supernova would be about a degree, whereas from his measurements its parallax could not be in excess of a few arc minutes, making the supernova at least ten times as far away as the Moon.

Five years later, Tycho was also able to show that the comet of 1577 had no parallax either, but traveled across the solar system, passing unhindered through any supposed crystal spheres that carried the planets, and proving that further dramatic changes took place above the Moon's orbit.

Other astronomers than Brahe had suspected that the supernova of 1572 had no parallax and must be among the fixed stars. Brahe's accurate measurements proved it, and that the celestial regions above the Moon were not as unchanging as Aristotle had thought. But Brahe was unable to bring himself to state this conclusion in his book *De Nova Stella*. On the contrary, he writes that all philosophers agree "that in the ethereal region of the celestial world no change of generation or of corruption occurs ... but celestial bodies always remain the same, like unto themselves in every way." He argued that God had concealed the supernova from earthly eyes since the creation, choosing to reveal it when He wished.

Brahe's cautious conservatism, which may have stood him in good stead in making careful observations, perhaps made him too unimaginative to perceive their consequences. He interpreted his observations of planetary motion in terms of epicycles and of his supernova in terms of suddenly revealed immutability. His pupil Kepler overthrew both concepts.

Kepler's Supernova of 1604

Tycho Brahe died in 1601, and his work was continued and developed by his former assistant, Johannes Kepler, a German. When the next supernova appeared in 1604, Kepler was working in Prague as court mathematician and tame astrologer for the erratic, and probably mad, Holy Roman Emperor Rudolf II.

In September 1604, many eyes were turned towards the region of sky in which Mars and Jupiter were slowly drawing together—a sight which would attract the attention even of non-astronomers. The supernova appeared in the nearby constellation of Ophiuchus and, thanks to the conjunction, observers saw the supernova when it first appeared and when its brightness was still increasing. It is rare for a nova of any kind to be discovered before it is at its maximum light, because the increase from obscurity to full brilliance is so rapid.

On October 9, two Italians saw the new star. One was an anonymous physician in Calabria, who reported what he had seen to the astronomer Clavius in Rome. The other was the astronomer I. Altobelli in Vienna. On October 10, a court official in Prague, J. Brunowsky, caught a glimpse of the new star between clouds and notified Kepler.

Unfortunately, the weather in Prague was cloudy from then until October 17, but some observations were made from places where the skies were clear. When Kepler did see the star on October 17, it was very striking. He wrote that it competed with Jupiter in brilliance and that

Johannes Kepler

it was colored like a diamond. This proved to be near the date of maximum brightness for the supernova as all observers agreed that there was
no further increase in brightness after October 15. Kepler made arrangements for continued observations to be made, but in November the
supernova was too close to the Sun to be seen. It reappeared from behind
the Sun in January 1605, by which time it was already fading. It continued to be visible until October 1605, and was carefully observed by
Kepler and others until then. When the Ophiuchus region came out
from behind the Sun in the spring once again, the supernova had become
invisible to the naked eye.

The European astronomers were not alone in observing the supernova.
It appears in Chinese and Korean records as well. Apart from their
interest in their own right, the Chinese observations are important
because they can be compared with the European results and enable us
to assess the accuracy of other Chinese records, such as those of the 1054
supernova. The 1604 star occurs as one of the last entries in a list of
guest stars in a document known as the *She-ke*. The entry is as follows:

*In the 32nd year of the same epoch, the 9th moon, Yih Chow [October 10,
1604] a star was seen in the degrees of the stellar division Wei. It resembled
a round ball. Its color was reddish yellow. It was seen in the southwest until
the 10th moon[October 27–November 26, 1604] when it was no longer visible.
In the 12th moon, day Sin Yew [February 3, 1605] it again appeared in the
southeast, in the stellar division Wei. The next year in the second moon
[March 24–April 23], it gradually faded away. In the 8th moon, day Ting
Maou [October 7th, 1605] it disappeared.*

This agrees entirely with the European records and indicates the
accuracy of the Chinese observations. Korean records also mention the
supernova, describing it as having a greater magnitude than Jupiter,
being a reddish-yellow color and scintillating. All European observers
in October 1604 remark upon the red, orange or yellow color.

Because of the philosophical controversies raging in 17th century

Europe one of the main reasons for the great interest which the supernova aroused was the question of whether or not the star, like Brahe's supernova 30 years before, was among the fixed stars. For this reason, great attention was given to measurement of its position in order to determine whether or not it had parallax. It had none large enough to be measured. Kepler produced a position for it but his figures are less satisfactory than those of David Fabricius at Osteel. From Fabricius's figures, modern astronomers have been able to compute the position to within one minute of arc.

The supernova caused great interest and much speculation. Kepler said that it would bring good fortune to publishers at least, as it would bring a spate of pamphlets and books. He himself at once rushed into print with an eight page pamphlet in German describing the star and comparing it with Tycho's star of 1572. In 1606 he published his book on the subject, *De Nova Stella*, which was, he said, "a book full of astronomical, physical, metaphysical, meteorological and astrological discussions, glorious and unusual." As this description makes clear, the book was by no means as astronomical paper as we understand the term, but it does contain astronomy. The Emperor Rudolf II's interest in astrology was such that Kepler was forced to include a great deal of interpretation and prediction. His main advice shows restraint: we should all consider what sins we have committed and pray for forgiveness.

Speculations about the supernova

The burning question to many, however, was this: was the star among the fixed stars and therefore another indication that there was change above the Moon? Kepler's answer was that, like Brahe's supernova, and for exactly the same reasons, the star was indeed above the Moon. Aristotle's model universe had failed in an important respect, and the rest of it was now under suspicion. Each of its attributes had to be subjected to scrutiny and tested against observation of the real world. Science became not a question of "saving the appearances," and making small modifications to an agreed-upon philosophy, but of attacking the basis of the philosophy itself.

Kepler went on to use Tycho's observations of the positions of the planets to determine that the orbits of the planets were not perfect circles at all, but flattened, forming ellipses. Aristotle's concept of perfection in the regions above the Moon was false in every detail.

Another controversial question concerned the origin of the star. One theory being discussed was that the conjunction of the planets had created such fire that they had ignited it. Kepler did not accept this idea. He believed that there is a material scattered through space which has an inherent ability to gather and then ignite itself. In favor of this idea, he argued that certain simple life forms appear spontaneously, in line with the common belief at the time that maggots appeared spontaneously in dead flesh. Shakespeare's Hamlet refers to this belief: "For if the sun breed maggots in a dead dog . . . ." He implies that Ophelia may conceive (though not necessarily spontaneously). Although Hamlet uses the idea ironically, it was a serious belief as Kepler's argument shows.

Controversy over the supernova was not confined to Prague where Kepler was working. In Italy, at Padua, Galileo Galilei was professor of mathematics at the university and had won a reputation for his brilliant lectures and treatises on mathematics and mechanics. Galileo's interest in astronomy, however, did not come to the fore until after 1609 when he made his first telescope. Nevertheless, in 1604 he was acquainted with the problems of astronomy. Tycho Brahe himself had written to Galileo in 1600 and invited him to enter into a scientific correspondence. Galileo seems to have snubbed Brahe and he certainly never entered into extensive correspondence with him. Later, Galileo showed himself extremely hostile to Brahe's ideas, particularly his system of the Universe.

When excitement arose over the 1604 supernova, Galileo, as a leading scientist, was asked to make a statement just as modern Nobel prize-winners are invited by the media to comment on scientific discoveries of which they often know little. Galileo was subjected to some criticism by his sponsors, the Padua city council, because he had not discovered the supernova, to which he somewhat peevishly replied that he had more important things to do than gaze out of the window, on the off-chance that he might see something interesting. Galileo was particularly pressed on the question of whether the star was superlunary or merely a meteorological phenomenon as the Aristotelians claimed.

Galileo seems to have been reluctant to make any statement. One reason for his scientific stature was that he required evidence from first hand observations before he would offer interpretations. He expressed his attitude clearly in a parable about a man who understood well the technicalities of how to produce musical notes, but when he held a singing cricket in his hand had no idea how it produced its song. Galileo went on:

The less people know and understand about such matters the more positively do they attempt to reason about them, and on the other hand, the number of things known and understood renders them more cautious in passing judgement about anything new.

When he did give three lectures on the supernova to overflow crowds in the largest hall in Padua, as far as is known he made no definite statement on one side or the other. No texts for the lectures have survived, but from the indirect evidence available he seems merely to have stated the case that had been so far made by each side, and he declined to draw conclusions. Apparently he began a book on the subject, but he never published it and only a small fraction of the manuscript remains. For the time being, he left the issue in doubt.

Explosions in the minds of men

The philosophical conclusions initiated by the supernovae of 1572 and 1604 and fully confirmed by Kepler's planetary theories, revolutionized the science of astronomy. But the new stars exploded in

43

the minds of non-astronomers as well as in the sky. The first attempts to understand them were astrological. Brahe speculated that a period of peace to be followed by years of violence was indicated by the clear, white light of the star of 1572 which had been followed by a red, martial light. There were many similar predictions. Theodore Beza, a French Protestant theologian, wrote a Latin poem suggesting that the star was the Star of Bethlehem, implying that it heralded the Second Coming of Christ. Queen Elizabeth I of England sent for a leading astrologer, Thomas Allen, to ask the meaning of the star, to which "he gave his opinion very learnedly." But as it had been established early on by Mästlin and Brahe that the star was not a comet, astrologers were being confronted with a phenomenon for which there was no clear European precedent. Sir Thomas Browne considered that the whole business of predictions by astrology had been brought into disrepute:

We need not be appalled by Blazing Stars, and a Comet is no more ground for Astrological presages than a flaming chimney.

From many writers of the 17th century, we receive an impression of doubt and uncertainty. Astronomers themselves were disputing the implications of the lack of significant parallax of the supernovae; what was a layman to think? Richard Corbet, Bishop of Oxford, in his letter to Master Ailesbury, written in 1618, expresses this uncertainty:

> *O tell us what to trust to; ere we wax*
> *All stiff and stupid with this Paralax.*
> *Say shall the old Philosophy be true,*
> *Or doth He ride above the Moon think you?*

If Corbet sounds unenthusiastic for the new problems of parallax, Henry More, a Cambridge scholar, writing later in 1642 is sceptical and uncomplimentary to those who still held to the old philosophy:

That famous star nail'd down in Cassiope
How was it hammer'd in your solid sky?
What pinsers pull'd it out again that we
No longer see it, whither did it fly?

Probably the most famous of the expressions of discomfort at the new ideas is from the 17th century poet, John Donne. His poem, *An Anatomie of the World* shows clearly how the discovery of supernovae could help to bring the structure of his whole world tumbling about a man's ears.

And new philosophy calls all in doubt ...
And freely men confesse that this world's spent,
When in the planets and the firmament
They seeke so many new.

Here, as in many of his comments on astronomy, Donne is accusing in his tone, implying that astronomers are forcing new worlds on our attention with all the new ideas that they imply, rather than merely observing them and reporting on what they observe. Donne points out the full logical conclusion which must be drawn from abandoning the old concept of the Universe. If the order in the heavens was gone, what about order in the state, the church, the family?

Tis all in peeces all cohaerence gone;
All just supply and all Relation;
Prince, Subject, Father, Sonne, are things forgot?

Reluctance to abandon all the old certainties is not surprising. At the same time, the "new stars" appealed to an age of explorers. There was some fitness that when the map of the Earth was rapidly being drawn larger and larger, there should be greater knowledge of the sky. Unlike Brahe, Donne implied that seeing a supernova is discovering a new star, rather than seeing an event which by its nature could not have been seen before. This attitude is part of his general hostility at this point to the "new philosophy." Likening America to the supernovae he thought that both would have the same effect on men's minds.

> *We have added to the world Virginia and sent*
> *Two new starres lately to the firmament.*

He addresses astronomers and explorers in one breath,

> *You, which beyond that heaven which was most high*
> *Have found new spheares and of new lands can write ...*

To Donne personally the new knowledge was cause for regret. There is no mistaking the poignancy in speaking of "heaven which *was* most high." Yet his frequent references to "new starres" show the impact that astronomy at that time would have on a man who thought about its implications.

For most people, the new stars co-existed peacefully beside their religion. To the compartmentalized or unquestioning mind, there were no uncomfortable religious or philosophical doubts. A supernova was merely a more than usually beautiful star. The poet Edmund Spenser tells a new bride to shine like a supernova.

> *Bee thou a new starre that to us pertends*
> *Ends of much wonder.*

John Dryden used Tycho himself in a poem as a means of praising Lord Hastings.

> *Liv'd Tycho now, struck with this Ray (which shone*
> *More bright i' the Moon than others' Beam at Noon)*
> *He'd take his Astrolabe and seek out here*
> *What new Star 'twas did gild our Hemisphere.*

There are no philosophical questionings, simply an acceptance of the supernova as a beautiful phenomenon which was part of the educated man's horizon at the time.

Tycho himself had compared the observation of the 1572 supernova with the stopping of the sun by Joshua or the Crucifixion, in its momentous effect on the mind and imagination. One feels that he would have been pleased by the wealth of writing in which his supernova was celebrated.

The questions which the two stars of 1572 and 1604 had raised, however, were to be answered finally and decisively only after the development of a new technology: the telescope.

Answers from the telescope

Historians are still arguing about who invented the telescope. Some say it was a Dutch optician, Hans Lipperhey, around 1608, while others maintain that the instrument had already been known for a decade or more before Lipperhey was granted a license to manufacture telescopes. Similarly, there is considerable evidence that Galileo was not the only person to turn a telescope on the stars in 1609, but what is certain is that it was Galileo who had the power of intellect to understand the implications of what he was seeing, and to relate that to the recent supernovae.

After experiments in producing several prototypes, Galileo designed a satisfactory telescope which gave him a clear view of the sky. Naturally, he pointed his telescope to well-known objects so that he could reveal new aspects of them. He found craters on the Moon, satellites orbiting Jupiter, rings around Saturn, spots on the Sun, new stars in the Pleiades star cluster ... the list goes on and on. His discoveries revolutionized astronomy.

He recorded his discoveries in *The Starry Messenger* (1610). Between 1623 and 1631 Galileo summarized his cosmology in his *Dialogue concerning Two World Systems*. The book is written as if it were a three way

47

conversation between an Aristotelian and a Copernican philosopher who try to convince an uncommitted disputant of the truth of their cause. Galileo puts his own view on the Universe through the arguments of the Copernican, Salviati, but he had been forced by the Pope to promise to give equal weight to the official Catholic position as well. He did this through the mouth of a character unsubtly named Simplicius and it is fairly obvious that Galileo's sympathies lay with the other debater, Salviati. One of the characters asks how astronomers could tell whether the new stars were "very remote." To this Salviati replies,

Either of two sorts of observations, both very simple, easy and correct, would be enough to assure them of the star being located in the firmament, or at least a long way beyond the Moon. One of these is the equality—or very slight disparity—of its distances from the pole when at its lowest point on the meridian and at its highest [the measurements made by Brahe, and described on p. 36]. The other is that it remained always at the same distance from certain surrounding fixed stars; especially Kappa Cassiopeiae, from which it [the 1572 supernova] was less than one and one half degree distant. From these two things it may unquestionably be deduced that parallax was either entirely lacking or was so small that the most cursory calculation proves the star to have been a great distance from the Earth.

The weight of evidence was now such that, as Galileo says, if Aristotle had been alive he would have changed his mind about the immutability of the heavens. Spots had been seen on the face of the Sun. Comets have been observed which have been

generated and dissolved in parts higher than the Lunar Orb, besides the two new stars, Anno 1572 and Anno 1604, without contradiction much higher than all the planets.

Nevertheless, Galileo had to pay the high price of a summons by the Inquisition and a trial in Rome for stating these views, which were

Galileo

unacceptable because they obliterated the distinction between the corruptible and incorruptible, placing the Earth among the heavens and bringing the heavens down to Earth.

Galileo suffered because his discoveries with his telescope made it impossible to maintain a belief in the unchanging heavens. He found himself obliged to state the conclusions which had in fact been inevitable since 1572. The supernovae had changed European cosmology.

What they mean to 20th century cosmology is equally momentous but fortunately less likely to arouse animosity or fear.

IV   *Supernovae in other galaxies: supernovae sought*

There has been, on average, a bright supernova visible in our Galaxy every 200 years. Unfortunately, not one has been seen since Kepler's in 1604. Astronomers have had to move gradually towards a better understanding of what supernovae are from observations of supernovae in other galaxies. When extragalactic supernovae were first seen they also caused incredulity and controversy, and when the modern founder of supernova research, Fritz Zwicky, began his systematic search for supernovae in other galaxies he was accompanied, he says, "by the hilarious laughter of most professional astronomers and my colleagues at Caltech." But before the first supernova was found by deliberate search, the understanding of supernovae continued to grow, helped by various chance discoveries.

A highly remarkable change

On the evening of August 20, 1885, E. Hartwig of the Dorpat Observatory in Russia was discussing the Laplace theory of the origin of the planets with friends. In broad outline this theory, put forward by Pierre-Simon Laplace in 1796, is similar to that widely accepted today.

The starting point of the theory is a huge slowly-rotating gas cloud.

As the cloud contracts its rotation speeds up until it becomes fast enough to throw off rings of material which then condense into planets. The central part of the cloud eventually becomes the Sun.

Around 1885, when Hartwig's philosophical discussion took place, astronomers were trying to link Laplace's theory with the observational fact that many nebulae were being discovered to have a spiral shape, though none had been resolved into stars. It was not surprising that many people thought them to be planetary systems in formation within our own Galaxy.

As Agnes Clerke, the British astronomer-writer, wrote with rash certainty in 1890: "No competent thinker, with the whole of the available evidence before him, can now, it is safe to say, maintain any single nebula to be a star system of coordinate rank within the Milky Way."

Hartwig's friends must have been eager to see one of these mysterious nebulae through a telescope, because he took them out to look at what was then called the Great Nebula in Andromeda, Messier 31. Through Dorpat Observatory's 9-inch refractor very little of interest would then have been visible, since the Moon was approaching Full with its bright light swamping the delicate structure in the nebula. Hartwig had surveyed M31 on three occasions during the previous New Moon period, and must therefore have been astounded to see at the center of M31 a new star shining brightly where no star had shone before. Clearly, thought Hartwig, this was a central sun appearing as predicted by Laplace's theory.

The star was seen in Hungary by Baroness Podmaniczky on August 22 or 23. She failed to realize its importance. At Heidelberg it was seen by Max Wolf on the 25th and 27th while testing a telescope, but he thought the star was an effect of moonlight. In Rouen, Ludovic Gully at a public night at the newly opened observatory on August 17 had seen the star in the new coudé telescope, but the telescope had been giving trouble in its tests and Gully thought that the appearance of the star was caused by a defect.

51

Only Hartwig saw the significance of the new star, presumably because he had been studying the region just the week previously. He could not convince the observatory's director of the new star's reality, however, and was not allowed to telegraph the discovery to the central clearing house for astronomical information in Kiel. Hartwig wrote to Kiel anyway but the letters went astray because of the petty theft of the stamps from the envelopes. He was not allowed to announce the "highly remarkable change in the Great Andromeda Nebula" until he and the director had confirmed the existence of the new star in the moonless sky on August 27. By that time the star had already noticeably faded. Hartwig was able to follow it almost daily as it declined. 180 days after its maximum, which had probably been on August 17, it was beyond the largest telescope's light-gathering power.

The nova was used in renewed attempts to decide whether or not M31 was a galaxy or a nebula. It had become clear that astrophysical arguments were too inconclusive because theories of the behavior of stars and interstellar gas were not well enough developed. For example, in 1899 J. Scheiner found that M31 had the spectrum of a collection of sunlike stars rather than the spectrum to be found in hot gaseous bodies. Scheiner correctly inferred that M31 was an aggregate of distant stars. But when V. M. Slipher examined the gas in the Pleiades star cluster, he found that it too had a starlike spectrum, and thought that the Andromeda Nebula might be the same. We now know that the Pleiades gas is dust-laden and simply reflects the light of the bright Pleiades stars.

Clearly only astronomical fact could determine the truth, and the prime fact required was the distance of M31. If M31 was no more distant than the stars of the Milky Way, it could not, in the picturesque phrase of the time, be an "island universe" of stars.

In the determination of the distance of M31 the nova observed by Hartwig, and now named S Andromedae, played a highly misleading part.

A misleading comparison

In 1911 F. W. Very compared S Andromedae with a nova which occurred in the Milky Way in Perseus in 1901. Nova Persei 1901 brightened in 28 hours from invisibility to naked eye brightness. Three days later, when it was at its brightest, it was among the half dozen brightest stars. It then faded over a few months back to invisibility.

Within six months Max Wolf noticed on photographs of the region of the nova that it had become surrounded by a small nebula. Because this was so soon after the nova outburst, astronomers realized that this nebula could not be gas ejected from the nova. Instead, this must be a reflection nebula. As the burst of light from the star spread at the speed of light into the surrounding space it illuminated a hitherto invisible dark nebula, reflecting light from its successive layers. As the nebula expanded, its brightness diminished: after two years it was a dim patch larger than the Moon and it gradually faded away.

Comparing the size of the nebula at various times with the speed of light at which it had been formed, astronomers were able to determine the distance to Nova Persei 1901 as some 500 light years.

At its maximum, therefore, Nova Persei was about 250 times brighter than S Andromedae at its maximum. Because the dimming of the light of distant objects depends on their distance squared, it followed that if Nova Persei had been moved about 16 times farther away it would have appeared to be 250 times fainter, the same as S Andromedae. Therefore, if S Andromedae and Nova Persei were alike, the Andromeda Nebula could be no farther than 8000 light years from Earth, not a very large distance and still well within our own Milky Way. From this, F. W. Very argued that M31 was not another galaxy like our own.

The discovery of further novae in spiral nebulae rekindled interest in this argument in 1917. G. W. Ritchey had photographed a nova in another spiral, catalog number NGC 6946, and this inspired him to re-examine all the photographs of spiral nebulae, including M31, taken by the Mount Wilson 60-inch telescope since 1908. Among the six

53

novae which he found were two in M31 that had passed unnoticed because they were much fainter than S Andromedae. Repeated photography of M31 by Ritchey, Harlow Shapley, John Duncan and R. F. Sanford quickly threw up eight further examples in two years, but none nearly as bright as S Andromedae.

It began to be clear that S Andromedae was not a typical nova in M31, and that, in making comparisons between novae in our Milky Way and in the Andromeda spiral nebula, S Andromedae should be ignored. Realizing instantly that the "occurrence of these new stars in spirals must be regarded as having a very definite bearing on the 'island universe' theory of the constitution of the spiral nebulae," H. D. Curtis went on to point out that the difference of brightness between novae occurring in our Milky Way and those in spiral nebulae, particularly M31, if S Andromedae were ignored, could be accounted for if the novae in M31 were 100 times the distance of Milky Way novae—"that is," writes Curtis, "the spirals containing the novae are far outside our stellar system." Curtis thus found himself a protagonist of the theory that spiral nebulae were galaxies like our own, "island universes" independent and separated from our Milky Way.

The Great Debate begins

In the same issue of the *Publications of the Astronomical Society of the Pacific* (October 1917) Harlow Shapley developed the same numerical argument as Curtis, but highlighted the problem of S Andromedae. He pointed out that it must have had a luminosity 100 million times that of the Sun, and that "this remarkable result must inevitably follow if spiral nebulae are considered external galactic systems comparable with our own in size and constituency."

During this period Shapley was studying the stars in dense spherical clusters (*globular clusters*), making a stellar census, and by 1919 had come to the conclusion that the study of globular clusters had yielded sufficient knowledge of the luminosity of more than a million stars to show that not one was anywhere near the enormous brightness of S Andromedae.

Hence, he argued, stellar luminosities of the order 100 million times that of the Sun seemed out of the question, and accordingly the close comparability of spirals containing such novae to our galaxy appeared inadmissable. Shapley found himself on the other side of the debate from Curtis, opposing the "island universe" hypothesis of the spiral nebulae.

A formal Great Debate on this subject was held on April 26, 1920 at the National Academy of Sciences in Washington, D.C. Curtis was forced to grasp the nettle of S Andromedae and to conclude that it seemed certain that the range of brightness of the novae in the spirals, and probably also in our Galaxy, may be very large, as is evidenced by a comparison of S Andromedae with the faint novae found in M31. A division into two classes was not impossible. With this statement Curtis ventured for the first time the concept that, besides ordinary novae, there exists a class of much brighter novae.

The Great Debate itself, of which the argument about the novae was a part, was inconclusive. Neither astronomer convinced the other, nor could their contemporaries decide what was the true status of the spirals. Not until 1923 did Edwin Hubble identify in M31 examples of a type of variable star called cepheids, using the greater power of the 100-inch Mount Wilson telescope to measure the varying brightness of these faint stars. Comparison of these stars with their Milky Way counterparts proved that, according to Hubble's figures, M31 was some 1 million light years away and was indeed a so-called "island universe" external to our own Galaxy.

Curtis' view of the status of the spirals was shown to be nearest the truth, and the Great Debate has been finally resolved in his favor.

Supernovae revealed

As the evidence began to be assimilated, the startling brightness of S Andromedae became more apparent. Since M31 is a galaxy like our own it contains stars by the billion. S Andromedae at maximum brightness was equal to one sixth of the light from the entire

55

galaxy! Edwin Hubble himself recognized in 1929 this consequence of his observations of the cepheids in M31: S Andromedae belongs to "that mysterious class of exceptional novae which attain luminosities that are respectable fractions of the total luminosity of the system in which they appear."

Up to 1933 haphazard photography of galaxies had thrown up a total of 19 examples of new stars having the property that they nearly equaled the brightness of the galaxies in which they were found. Walter Baade and Fritz Zwicky christened these stars *supernovae* and outlined their fundamental properties. To them belongs the credit for being the first to understand the key part that supernovae were to play in modern astronomy. In its entirety the summary of their paper, translated into non-technical language, reads as follows:

Supernovae flare up in every galaxy once in several centuries. The lifetime of a supernova is about 20 days and its brightness at maximum may be as high as 100 million times that of the Sun. Calculations indicate that the total radiation, visible and invisible, is about 10 million times what can be seen. The supernova therefore emits during its life a total energy equal to the amount that the Sun would radiate in a million years. If supernovae are initially quite ordinary stars of mass up to about 10 times that of the Sun, the amount of energy they release is comparable to the energy that would be made if their mass was turned directly to energy. Therefore in the supernova process mass in bulk is annihilated.

In addition, the hypothesis that cosmic rays are produced by supernovae suggests itself. Assuming that in every galaxy one supernova occurs every thousand years the intensity of cosmic rays expected to be observed on Earth is equal to the level actually observed. With all reserve we advance the view that supernovae represent the transitions from ordinary stars into neutron stars which in their final stages consist of extremely closely-packed neutrons.

In a single 15-line summary, fully aware of the bizarre nature of what they were saying, Baade and Zwicky mapped out the achievements of the next 40 years' work on supernovae, as the details of their outline were filled in.

Zwicky himself recalled that when in 1934 he bought a camera just to prove that he and Baade were right by photographing the rich cluster of galaxies towards the constellation Virgo from the top of a building of the California Institute of Technology he was scorned by his colleagues. Expecting to find two or three supernovae in two years he in fact found none. Though he was almost inclined to give up the systematic search for supernovae because of this setback, Zwicky persuaded George Ellery Hale, Director of the Mount Wilson Observatory, to divert some money from the grant from the Rockefeller Foundation to build the 200-inch telescope in order to construct a new type of camera which would better enable him to find supernovae in other galaxies.

Bernhard Schmidt, an Estonian optical designer, had recently suggested a way of giving an astronomical camera-telescope a much larger field of view, using a corrector lens. Instead of the usual half-degree or so, the new Schmidt camera photographed a circle of sky eight degrees across and had an aperture of 18 inches—an ideal instrument for looking at many galaxies at once. Zwicky got his telescope.

In the period from September 1936 to December 1939, Zwicky and his co-worker J. J. Johnson took 1625 photographs of 175 regions of the sky chosen to contain nearby galaxies. Taking into account the number of galaxies in the field of the telescope the program amounted to 5150 years continuous observation of an average galaxy. Twelve supernovae were discovered in those three years, giving a rate of one discovered supernova every 430 years per average galaxy. The second which Zwicky discovered, at maximum brightness on August 22, 1937, was more than 100 times brighter than the total light of the galaxy in which it appeared, an irregular spiral galaxy called IC 4182, and was the brightest supernova seen so far in this century.

During World War II the search essentially stopped, but Zwicky persuaded Hale to ask the Rockefeller Foundation for nearly $500,000 to construct a larger 48-inch Schmidt telescope, put into operation in

57

A supernova flares in a distant galaxy. The top picture shows the galaxy NGC 5253 as it usually appears on photos taken with the 48-inch Palomar Schmidt.

January 4, 1959

May 6, 1972

June 10, 1972

1949. This telescope's first job was to photograph the entire sky in the Palomar Sky Survey, which has been a cornerstone of astronomy for many years; but the Palomar Supernova Search was resumed in 1958 and is still under way, directed by Wallace Sargent and Leonard Searle. The man who carries out the photography and who makes the discoveries is Charles T. Kowal, who has compared himself to a ship's engineer, making the ship run while the captain walks the bridge. Three moonless nights per month are scheduled for the search, which is of 38 fields containing a total of 3003 galaxies in clusters and groups. The photographs taken at night are compared the following afternoon, if possible, with standard photographs of the same fields, so that newly discovered supernovae can be rapidly followed up. Between 1958 and 1971, the search produced 63 supernovae, and the total discovered at Palomar

The lower four show the 1972 supernova in this galaxy. The brightness of the supernova declined from its maximum (far left), when it was as bright as all the other stars in NGC 5253 combined, to near-invisibility after almost a year (far right). The fifth supernova to be discovered in an external galaxy in that year, 1972e was of Type I. It lies well from the center of its galaxy, in regions too faint to be recorded on these photographs. Hale Observatory photographs by Charles T. Kowal.

January 30, 1973

April 24, 1973

since the inception of Zwicky's search up to 1973 was 270 supernovae, at a cost, noted Zwicky, of $550 each. Other supernova searches are carried out from Asiago (Italy), Zimmerwald (Switzerland) and Konkoly (Hungary).

The supernova birth rate

It is difficult to determine the true rate at which supernovae occur in the galaxies searched. The statistics show that fewer supernovae are found in the fainter galaxies than in the brighter, presumably because the fainter galaxies are, on average, farther away, so supernovae in them will be correspondingly less noticeable. Supernovae are also missed when they merge with the central regions of the fainter galaxies.

59

Sargent, Searle and Kowal have concluded that in a typical galaxy the average supernova frequency may be as high as one every 20 to 30 years, considerably more than Zwicky's first estimate. Up to his death in February 1974, Zwicky himself vigorously disputed such a high frequency.

This high rate of occurrence of supernovae, even in our own Galaxy, is to some degree confirmed by studies of the number of pulsars. Radio astronomer Andrew Lyne has offered evidence based on a Jodrell Bank pulsar survey that the number of pulsars now visible suggests that they are created in our Galaxy at the rate of one every 10 to 20 years. If, as astronomers believe, pulsars are created in supernova explosions, the pulsar birth rate and the supernova birth rate should be the same.

Where have all the supernovae gone?

The Sun lies almost exactly in the plane of the Galaxy— along the Galaxy's equator in effect—which is where supernovae mostly occur. If they occur as often as one every 20 years in our Galaxy why don't we see more? Why is it that astronomers have waited 300 years without seeing a galactic supernova? The culprit is the dust and other interstellar material which also lies along the galactic plane, obscuring much of what goes on even in our own neighborhood. Look out at the Milky Way on a dark moonless night. In Cygnus and Taurus the Milky Way appears cleft in two. In the Southern Cross a dark hole is silhouetted against the Milky Way. All these black patches are relatively near dark clouds of dust hiding the light of the stars behind. Calculations show that only 40 per cent of all supernovae that occur in the Galaxy can be visible to the naked eye, even at maximum. Apparently, over the last 1000 years, only ten per cent have reached a sufficient brightness for a long enough time, far enough from the Sun for their light not to be swamped by the dawn or evening sky, to make discovery possible. Modern astronomers cannot accept a 90 per cent failure rate, and have made special arrangements to catch the next galactic supernova.

The Milky Way from the constellation of Scutum (top), to Scorpio at bottom. The horizon is blurred as the telescope taking this picture followed the stars. Silhouetted against the massed faint stars of the Milky Way are lanes of dark clouds which blot out the light from stars behind. Galactic supernovae preferentially occur along the center line of the Milky Way and are heavily obscured by the dark clouds.

Looking for the next galactic supernova

The last time anybody was able to stare up at the stars and see a supernova was in 1604. Despite the statistics, there has not been one visible to the unaided eye for almost 400 years. Obviously, the next time the Earth is illuminated by the glare of such a star, the attention of all astronomers will be focused on it. But whenever the next galactic supernova appears—and it could be tomorrow—the first problem will be to recognize it and distinguish it from ordinary novae.

Ordinary novae in our Galaxy are seen to flare up at the rate of one or two per year. They are often picked out by watchful amateur astronomers who make a point of searching for them. The sky is so large, and nova searching so frequently fruitless, that professionals can find little time for this sort of thing. Amateurs, however, have the time and dedication to learn the appearance of the sky well beyond the limit of naked-eye vision. The game can be rewarding because, as with comets, new stars are sometimes named after their discoverer, who also has the satisfaction of knowing that the world's major telescopes will be turned towards his star.

As a result of the keenness of the amateur nova patrol, many novae are now being spotted when they are too faint to be seen with the naked eye. This greatly increases the chance that even if a distant, heavily obscured supernova were to appear in the present day, it would still be spotted.

Because the occurrence of such things as supernovae cannot be predicted and because they brighten up so quickly, astronomers have now set up an early warning system, run by the International Astronomical Union. Anyone—professional or amateur—discovering an important new object such as a nova or comet notifies the Central Bureau for Astronomical Telegrams in Cambridge, Mass., usually through the facilities of the nearest observatory. The Bureau then cables it subscribers, which include all principal observatories and amateur groups, using a brief and economical code, prefaced by the alerting codenames ASTROGRAM ECHO (for bright novae) or ASTROGRAM FRANCE (for fainter ones).

When an observatory receives one of the innocuous-looking strings of five-digit numbers, there is always a flurry of activity as astronomers decipher the groups and check charts of the right area of sky in their libraries. They then telephone details to colleagues at their observatory's telescopes. It is reasonably safe to predict that every suitable telescope will be pointing at it within 24 hours of the discovery of the next galactic supernova.

If the next supernova occurs in the direction away from the center of the Galaxy, then it will almost certainly be bright enough to be visible to the naked eye. Most supernovae, however, are expected to lie towards

After an initial brief telegram, the world's astronomers are given details of important discoveries by an I.A.U. Circular such as this one, airmailed rapidly to thousands of professional and amateur subscribers. This Circular, sent out May 18, 1972, announces Kowal's discovery of the supernova shown on pp. 58–59.

Circular No. 2405

CENTRAL BUREAU FOR ASTRONOMICAL TELEGRAMS

INTERNATIONAL ASTRONOMICAL UNION

POSTAL ADDRESS: CENTRAL BUREAU FOR ASTRONOMICAL TELEGRAMS, SMITHSONIAN ASTROPHYSICAL OBSERVATORY, CAMBRIDGE, MASS. 02138, USA

CABLE ADDRESS: SATELLITES, NEWYORK - WESTERN UNION: RAPID SATELLITE CAMBMASS

SUPERNOVA IN NGC 5253

Mr. C. T. Kowal, Department of Astrophysics, California Institute of Technology, telegraphs that he discovered on May 13 a supernova of magnitude 8.5 in NGC 5253 ($\alpha = 13^h37^m1$, $\delta = -31°24'$, equinox 1950.0). The object, located 56" west and 85" south of the nucleus, was confirmed on May 15. This seems to be the fourth brightest extragalactic supernova ever recorded; the second brightest (= Z Cen), observed in 1895, was also in NGC 5253.

R CORONAE BOREALIS

Recent observations show that this object has been brightening again. C. E. Scovil, Stamford, Connecticut, gives the following magnitude estimates: May 5.17, 11.2; 6.24, 11.3; 7.11, 10.5; 11.28, 9.6; 12.14, 9.6; 13.28, 9.7; 14.18, 9.6. P. Moore, Selsey, Sussex, England, gives: Apr. 29, 12.1; May 7, 10.7; May 10, 10.4.

PERIODIC COMET KEARNS-KWEE (1971c)

In calculating the following ephemeris (cf. IAUC 2330) the ΔT correction (as indicated by several semi-accurate observations by Dr. E. Roemer during July-December 1971) has been applied.

1972/73 ET	α_{1950}	δ_{1950}	Δ	r	m_2
June 12	2^h40^m98	+22°44!8	3.336	2.591	18.3
22	2 59.65	+24 16.9			
July 2	3 18.70	+25 43.7	3.111	2.518	18.0
12	3 38.11	+27 04.3			
22	3 57.77	+28 18.0	2.867	2.451	17.7
Aug. 1	4 17.60	+29 24.2			
11	4 37.46	+30 22.2	2.612	2.391	17.4
21	4 57.16	+31 11.8			
31	5 16.50	+31 53.1	2.353	2.340	17.0
Sept. 10	5 35.23	+32 26.3			
20	5 53.01	+32 52.1	2.095	2.297	16.7
30	6 09.54	+33 11.5			
Oct. 10	6 24.39	+33 25.6	1.848	2.264	16.4
20	6 37.14	+33 36.0			
30	6 47.33	+33 43.6	1.622	2.241	16.1
Nov. 9	6 54.48	+33 49.2			
19	6 58.21	+33 52.6	1.434	2.230	15.8

the galactic center, just because most of the Galaxy as seen from the solar system lies in that direction. If the supernova is relatively close to us, say at a distance of a few thousand light years, it will be possible to see the supernova grow in size as it expands into space. Only 40 days after it explodes a supernova at a distance of 3000 light years will appear non-stellar in a telescope. Possibly, if some theories of supernovae are correct, it will be large enough to be seen as a perceptible disk right at the time of maximum brightness even when observed by amateur astronomers with moderate-sized telescopes. To the naked eye the supernova may appear not to twinkle to the same degree as other bright stars, but might glow balefully.

The burst of light from the supernova as it explodes will spread into the surrounding space, at the speed of light, and will illuminate any interstellar material that happens to lie near the supernova as was the case with Nova Persei 1901. Light from the supernova explosion will be reflected towards the Earth by previously unseen dark clouds of interstellar material and a ripple of light will be seen spreading out through interstellar material, away from the supernova. These light echoes may be visible for hundreds of years, expanding into space until the light is too diffuse to be seen. If they are visible to the naked eye, the light echoes will probably be detectable to a distance of one degree from the supernova, and the light echo will appear as a ring about four times the size of the Moon.

If such a thing is seen surrounding the next galactic supernova, it will provide an easy way to determine its distance. One year after maximum, the radius of the light echo will be exactly one light year. Astronomers will be able to measure its angular distance at this time from the supernova and determine, by trigonometry, the precise distance of the supernova.

Although supernovae emit large numbers of X-rays and gamma rays, this high energy radiation will be prevented from reaching Earth by the dense expanding shell ejected by the explosion. There is possibly a brief moment at the beginning of a supernova explosion when there is some hope of detecting a blast of X-radiation, the so-called *prompt emission*, if there is an X-ray satellite orbiting the Earth at the time. A few years after a supernova explosion, the shell becomes thin enough to be transparent to X-rays again, and they could probably be detected even then if they are strong enough.

As the shell ejected by a supernova disperses into space, it may cool to a point where dust grains can form. These will be warm when they first form and will emit infrared radiation. Therefore, at about the time when X-radiation becomes visible, infrared radiation will also be perceptible, while the supernova optically fades away.

These ideas about the next galactic supernova are all based on a discussion presented by Sidney Van den Bergh in a paper intended to help astronomers to plan observations, should the event occur in their working lifetime. Van den Bergh remarks at the end of his article: "Those who might tend to become discouraged while they wait for this momentous occasion might be slightly consoled by the thought that the light of about 500 galactic supernovae that have already occurred is currently on its way to us!"

V *The Crab and its mysteries: a supernova remnant*

While waiting for the next bright supernova to study, astronomers have been studying the remnants of past supernovae, for their own interest and for the light they throw on the supernova phenomenon. Probably more effort has been put into understanding one particular remnant, the Crab Nebula, than any other astronomical object, save the Sun. Solving the mysteries of the Crab Nebula has progressed with the development of new astronomical instruments and techniques, starting with the invention of the telescope itself. Gradually, astronomers have come to understand how significant the tantalizing and infuriating Crab could be in understanding supernovae. No matter how many of its mysteries have been solved, the Crab seems always to hint at another problem.

Like all the first-discovered nebulae, it was found by chance, not through a systematic search, because the invention of the telescope had posed astronomers a question which most, like Galileo, declined to tackle. How could they systematically survey the whole sky with the telescope and still have time for other studies? Suppose we estimate the field of view of his telescope—the area the astronomer could see at any one time—at one degree (twice the diameter of the Moon). This figure is generous for telescopes of the 17th and 18th centuries. Over the whole sky there are 42,000 such areas of which perhaps a quarter are perpetually below the horizon from European latitudes. Allowing just four minutes for the inspection of such an area as the telescope, carried by the Earth's rotation, scans across it, we find that some 2000 hours are needed to sweep the telescope over the whole sky. Astronomical telescopes view the sky at night for about half the time (the other half is spent setting up the telescope, making notes, and so on); it is cloudy in Europe for half to two thirds of all nights; and half the time the Moon is too bright and floods the sky with light, washing out the fainter stars. Consequently, to be able to observe for 2000 hours in the best conditions requires about seven years, assuming that the astronomer is dedicated and is prepared to work throughout the year during all hours of darkness. Realistically, such an all-sky survey takes a substantial fraction of a working lifetime. Yet as telescopes have improved, astronomers have repeated examinations of the sky even to the present day.

One such early astronomer was John Bevis (1693–1771), a Welsh doctor, who compiled the results of his observations of star positions into an atlas, the *Uranographia Britannica*. The costs of preparing the plates for the Atlas were so high that the printer went bankrupt and his creditors seized his assets including the engravings of the star charts. A few proofs of the *Uranographia Britannica* had been struck, however, and on these are plotted 16 *nebulae* (Latin for *clouds*), the term astronomers now use to refer to the unstarlike patches of light which they were beginning to discover.

One set of proofs was shown to the French historian Lalande on a visit to London. Lalande, in a text on astronomy, records that the great French astronomer Charles Messier had another set. In 1758, Messier, like the whole of the astronomical community, was eagerly awaiting the appearance of Halley's Comet. Many years before, Edmond Halley had realized that this bright comet, then last seen in 1683, returns to the vicinity of the Earth and Sun every 75 years or so. Messier was known as the "ferret of comets" for his assiduous searches for and discoveries of new comets. He actually found a comet near the predicted place, but in fact it turned out not to be Halley's, which arrived later. Messier's new comet passed into the constellation Taurus, and in following it he chanced upon Bevis' nebula. In his own words: "The comet of 1758 being between the horns of the Bull, I discovered on August 28 below the southern horn and a short distance from the star Zeta of that constellation, a whitish light, elongated in the form of a candle flame containing not one star." He compared the nebula with the comet and said that the nebula was more "vivid" and more elongated than the comet which seemed "almost round."

This discovery was the first of many nebulae found by Messier in the course of sweeping the sky for comets. In 1771 he published a catalog of all known nebulae with Bevis' nebula in Taurus in first place. After Bevis wrote to Messier pointing out that he had discovered the nebula first, Messier gave him the credit for it. Messier's catalog of nebulae is known by the initial of its compiler, and therefore Bevis' nebula is called M1.

Although it is first in the catalog, M1 is by no means the most prominent nebula in the sky. Others, such as the Orion Nebula (M42) are much more spectacular and can easily be seen with the naked eye. While M1 can be seen as a misty patch in fairly small telescopes under good conditions, it needs a moderately large instrument to show it well.

Stars or gas?

As telescopes became better and better, astronomers began to find that many of the 103 so-called "nebulae" in Messier's catalog were in fact clusters of faint stars packed so closely together that the individual stars could not be separated or *resolved* in poorer telescopes. Naturally, astronomers speculated whether all the nebulae would eventually be resolved if large enough telescopes could be trained on them.

For a long time astronomers believed that nebulae were "cosmical sandheaps too remote to be resolved into stars." In particular, Bevis' nebula, M1, always gave the impression when seen in better and better telescopes that it was just on the point of being resolved. William

Herschel inspected it many times in the course of what he called his "star-gaging" with the vast telescopes which he had made specially for sweeping the entire heavens. In 1818 he wrote, "it is resolvable" (but not, notice, "it is resolved") and went on: "There does not seem to be any milky nebulosity mixed with what I take to be small lucid points. As all the observers agree to call this object resolvable, it is probably a cluster of stars at no very great distance beyond my telescopes' gaging powers." His son, John Herschel, placed M1 first in a sequence of nebulae which had turned out to be clusters of stars, resolvable by successive degrees, presumably because he too could glimpse the "lucid points" noted by his father. In fact, M1 was said by John Herschel to be "hairy" or "filamentous."

The nebula was observed by the Earl of Rosse with his telescope of three-foot aperture. In 1844, writing on his observations of nebulae, Rosse wrote: "Now, as has always been the case, an increase of instrumental power had added to the number of clusters (of stars) at the expense of the nebulae, properly so called; still it would be very unsafe to conclude that such will always be the case." But in spite of Rosse's caution here, he too was of the opinion that M1 was a star cluster just beyond the resolution of his telescope as it had been to Herschel's smaller instruments: "It is studded with stars mixed, however, with a nebulosity probably consisting of stars too minute to be recognized."

Rosse added a novel aspect to the description of M1, calling it for the first time the Crab Nebula. He wrote that, with his telescope, "it is transformed to a closely-crowded cluster, with branches streaming off from the oval boundary, like claws, so as to give it an appearance that in a measure justifies the name by which it is distinguished." Rosse published a very crustacean-like picture of M1 in 1844 and it has been known as the Crab Nebula ever since.

Rosse created a memorable name for the nebula, but it is sad to say that its actual appearance does not live up to its imaginative title. His contemporary, Dreyer, said that Rosse's 1844 drawing was not at all like the real nebula, and 20 years later Rosse published a completely

British aristocrat Lord Rosse's first published picture of the Crab Nebula christened it with its distinctive name, but this representation looks nothing like Rosse's subsequent accurate drawings of the real nebula.

different drawing, made with his six-foot telescope, then the world's largest, which he erected near Birr Castle in Ireland. Slung between two towers, the telescope could move only up and down and had very limited ability to track stars as they crossed the north–south line through the telescope. With it, however, Rosse was able to show that many nebulae had a characteristic spiral shape, and we now know that these are distant galaxies mostly too far away for the individual stars to be seen. Although observing with his telescope must have been difficult, the fact that he was able to give such accurate impressions of spiral galaxies shows that what he and his co-workers saw with his larger telescope and what he drew had firm roots in reality. His later picture of M1 is in fact close to modern photographs.

The reason for his more fanciful earlier picture is not clear but others followed him in perceiving something strange about M1. William Lassell, an English amateur astronomer who was a brewer by profession, also remarked on the filaments and "claws" that he could see when he viewed M1 in 1853 from the clear skies of Malta. Lassell astutely noticed

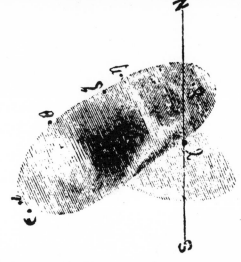

that he could see no more stars in it than were contained in an equal area
of other parts of the sky nearby, from which he inferred that the stars he
could see within M1 apparently had no connection with it. This marks
the beginning of the evidence that the Crab Nebula was not going to be
resolved into stars when a suitably large telescope was built, but would
remain cloudy and nebular.

The Crab Nebula was first photographed in 1892 by Isaac Roberts,
a leading astronomical photographer. Roberts was rather unfairly
dismissive about Rosse's 1844 descriptions and Lassell's 1853 drawings,

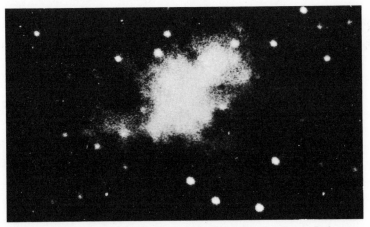

*The first photograph of the Crab Nebula, taken by Isaac Roberts,
clearly recorded the deep bay at the east (left) side and smaller
bays to the north (top) and west. Mottling due to the filaments
was reported to be visible on the original negative.*

to which he said his photograph showed no resemblance, though he acknowledged that Rosse's picture of 1861 had several features that approximately corresponded with the photograph. He described the nebula as elongated, irregular in outline with a deep bay on one side counterbalanced by a projection on the other. The original negative, he said, showed mottling, rifts, and some starlike condensations in the nebulosity. Apparently Roberts was being careful in saying "starlike" here; he seems unwilling to say that the condensations *are* stars, and his caution has since been justified.

So how does one discover whether the Crab Nebula is made of stars or gas? The first major step forward came when two astronomers at Harvard Observatory, Joseph Winlock and E. C. Pickering, studied the nebula through a device called a spectroscope.

The spectroscope ranks next to the telescope itself in its importance to astronomers, since it has the power to analyze the light from glowing bodies. It is worth spending some time explaining how it does this, and what the results mean.

It was Isaac Newton who first found that white light passed through a triangular glass prism is split into the colors of the rainbow—what scientists call the *spectrum*. In the early spectroscopes, light gathered by an astronomical telescope was split up by prisms and then examined visually with a small telescope.

In the first applications of the spectroscope to astronomy it was found that the spectrum of light was continuous—it spread into *all* the colors of the rainbow. A rainbow is in fact the spectrum of a star—the Sun— spread into its constituent colors by the prism-like action of raindrops. But William Huggins found in 1864 that a large number of nebulae showed not a continuous spectrum, but one in which only a small number of individual colors called spectral lines occurred. The explanation for this is that light from a star represents energy given out by the particles of which it is made. A single color represents a packet of energy of a certain size. If the object on which a spectroscope is focused is giving out packets of energy of all conceivable sizes, its spectrum

72

contains all colors. This is the case for the light from the dense parts of a star's atmosphere.

The star is composed, like the Earth, of individual particles of matter called atoms which are the basic building blocks of the chemical elements. Unlike atoms on Earth, however, the atoms in a star are hot enough to be moving rapidly back and forth, colliding with each other violently, and the force of the collision is enough to fragment the atoms into some of their constituent pieces. The outer bits of the atoms can be knocked free and these bits join in the jostling of the hot atoms. The bits are called electrons. Unless it has been fragmented each atom has a precise number of electrons, which makes it the kind of atom that it is. The central part of an atom, its nucleus, is positively charged, and electrons are negatively charged; therefore the atom is tied together by electrical forces. Only the violent collisions in hot gases at temperatures like those found in stars can overcome the electrical forces. The result is that a star's atmosphere consists of a dense gas of electrical particles, including electrons, rushing back and forth, jostling each other.

Now light of any color (and indeed radio waves and X-rays as well) is commonly produced by just one basic process, though that process can occur in a range of circumstances. Whenever a charged particle is decelerated (or the direction of its motion is changed, which amounts to the same thing), it radiates a pulse of energy, called a *photon*. The wavelength (color) of the energy depends partly on the violence of the change in motion. So in a star, the random jerks experienced by the electrons produce a whole range of wavelengths, and we see a continuous spectrum.

It is worth mentioning in passing that this continuous spectrum usually has a predominant color, depending on the temperature of the star. As objects get hotter they glow first red, then yellow, then white, then blue as their peak of energy shifts farther to the energetic violet end of the spectrum. The electrons in a hot star generally move faster than in a cool star, and the cool star therefore appears redder than the hot star.

73

In the same way, the crowd at a football match have more violent encounters with one another than an equally densely-packed crowd at, say, a chess tournament, even though a few individual members of either may be similarly excited or downcast. So there is a spread of energies within both crowds, just as there is a spread of colors emitted by stars of different temperatures, but one degree of excitedness predominates.

Suppose, having looked with a spectroscope at a dense gas such as exists in stars, we turn to inspect a more rarefied gas. This means that there are long distances between atoms, and encounters are few and far between. Little continuous light is given out. The light from such a gas is not the energy emitted from random encounters but from energy changes which take place among the electrons inside the atoms themselves. In a rarefied gas, the atoms have more chance of retaining their own electrons, whereas in the chaotic conditions within a star, the electrons are usually free to move around. All atoms of a given kind contain precisely the same energy levels, like a series of steps, and as an electron moves from one level to the next it gives out a step of energy of the same size as an electron undergoing the same step in any other, similar atom. All atoms undergoing this energy change give out light of the same color—a spectral line.

Thus a dense material gives out a continuous spectrum but only a rarefied gas can give out spectral lines. Although the full explanation was not known to Huggins, his own observations showed that since obvious star clusters always had continuous spectra, while unmistakably gaseous nebulae always showed bright individual spectral lines, the spectroscope offered a ready means of deciding uncertain cases. Many showed a green spectral line, unknown before its appearance in nebulae, and thought possibly to come from a kind of atom not present on Earth.

Indeed, a few years previously astronomers had found several strong lines in the Sun's spectrum which did not tally with any gas then known on Earth. The mystery gas was christened after *helios*, the Greek for "Sun," which is how helium got its name. Similarly, the new element in the nebulae was named nebulium. Later analysis showed that the green line was actually the oxygen atom in a state unfamiliar on Earth.

The light from the Crab

Huggins does not seem to have observed the Crab Nebula through his spectroscope but Winlock and Pickering at Harvard in 1868 saw the green nebulium spectral line, which proved that at least part of the Crab Nebula was gaseous.

They reported however that the spectrum showed also an unusually strong continuous component. This must have been puzzling as it suggested that there were more stars than usual embedded in the nebula. In fact, such an interpretation is wrong. The continuous spectrum of the Crab Nebula is caused in a quite different way from that of a star—it arises from the interaction between electrons and a strong magnetic field. This pale clue to a very important process in astrophysics was first noticed in particle accelerators called synchrotrons—hence its name of *synchrotron radiation*.

Modern color photographs show that the two spectral components of the Crab Nebula come from distinctly different structures. The nebula consists of a lace of red and green filaments, generally oval in total outline, embedded in and around a tenuous, milky-white light. Many of the filaments are green where the green "nebulium" spectral line predominates but filaments of a red color are common. The red spectral line comes from electrons dropping down the first energy step in atoms of hydrogen, and is called H-alpha. It was not seen visually, probably on account of its deep red color which occurs where the eye's color sensitivity is poor. Further, weaker spectral lines from other energy steps in the hydrogen atom, called H-beta, H-gamma and so on, are present in the spectrum of the filaments as well, shining green and blue. Other spectral lines from sulfur, helium and neon can be seen in a spectroscope.

Knowing that the nebula has this curious knotted appearance of

75

BLUE λ3100-λ5000

RED λ6300-λ67

Two faces of the Crab. Mt. Wilson astronomers filtered the light of the 100-inch telescope to restrict the wavelengths used in making this pair of photographs. Two components of the Crab's light are seen: blue light (left) shows a featureless glow while red light (right) picks out the lacy filaments of hydrogen gas.

filaments embedded in a smoother uncolored component, we can see now why the 18th and 19th century visual observers always had the impression that the Crab Nebula was about to be resolved into stars. Descriptions such as John Herschel's "hairy" referred to the filaments, while the knots fooled Rosse into believing that he had really glimpsed stars embedded in nebulosity. If he had called the stars "lucid points" or "starlike" (as did William Herschel and Isaac Roberts respectively) he would have given the most accurate visual description of all.

Motions in the Crab Nebula

The mere existence of spectral lines in the spectrum of the Crab Nebula gave astronomers the vital clue that they needed to decide that it was made not of stars but gas. Measurement of the precise wavelengths of the lines would enable them to study the motions of the gas in the Crab. Nineteenth century astronomers understood the principle of how this would be accomplished but could not marry theory with practice until the invention of the photographic plate and its application to astronomical spectroscopy. The principle they hoped to exploit was the "Doppler effect."

Doppler shifted violinists

The Doppler effect is a shift of star's spectral line because of the star's motion towards or away from Earth. It was first explained in 1842 by Christian Doppler, an Austrian physicist, with reference to sound waves rather than light waves. For one experiment, he hired a group of violinists, who had the gift of absolute pitch so that they could tell precisely what note an instrument was playing without reference to any external standard. Doppler asked the musicians to sit in an open railroad car and play one particular note as the car moved at various speeds along a length of track. While they did this, another musician stood by the track and listened to the note that the violinists were playing. When the car was approaching the stationary violinist, he heard a higher note than the one actually being played. When the car was receding, he heard a lower note. Most people, standing at a railroad station, have experienced the difference in sound between an approaching train and one that is receding.

There is close correspondence between what happens in the case of sound waves, and what happens in the case of light waves. We talk about colors or spectral lines instead of musical notes but the two concepts are identical. Instead of a "higher note," we talk about a "blue-shifted spectral line," and instead of a "lower note," a "red-shifted spectral line." The degree to which a spectral line is moved towards the blue or red end of the spectrum is a measure of how fast the object which emitted this light is moving towards or away from us. No matter how distant the star, as long as spectral features can be distinguished

77

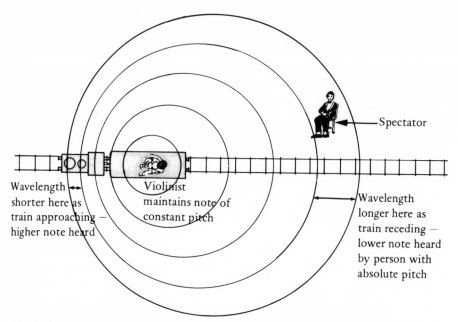

Wavelength
shorter here as
train approaching —
higher note heard

Violinist
maintains note of
constant pitch

Spectator

Wavelength
longer here as
train receding —
lower note heard
by person with
absolute pitch

*In C. Doppler's experiment successive waves of sound were emitted by the
violinist as he moved along the track on the flatbed car. The waves were
crowded together in the direction of motion of the train, spread apart to rear of
train.*

astronomers can measure its speed along the line of sight. If the star is
fairly close, it will also change its position in the sky over a period of
time and so its motion in all directions is known.

In practice the Doppler shift in a typical astronomical object is small.
The wavelength of its spectral lines might change, typically, by 0.03%.
Astronomers' first attempts to measure Doppler shifts were foiled by
their smallness. A spectrum had to be examined in fine detail before
they could be perceived. Of course the finer the detail in which you need
to look at anything the more light you need to be able to see it, just as to
read small print you need a brighter light than to read a newspaper
headline. The eye could not grasp enough light to "read" fine Doppler
shifts in the first spectroscopes. Only when the astronomer's eye was
replaced by a photographic plate to record the spectrum was it possible

to measure the Doppler shift caused by motions of stars and nebulae.

The photographic plate can do what the eye alone cannot: it integrates. That is to say, unlike the eye which perceives a new image about every $\frac{1}{25}$ second, a photographic plate can be put at the focus of a spectroscope for many hours to store up, or integrate, the photons that the sky is sending. Since the spectrum appeared as a permanent picture rather than a fleeting image in a man's eye, the spectroscope was renamed the "spectrograph." With a spectrograph it became possible to record the wavelengths of light and the way in which light was distributed among the various wavelengths in finer and finer detail.

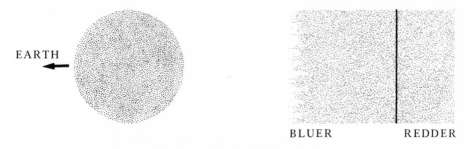

EARTH

BLUER   REDDER

A stationary nebula gives single emission lines in its spectrum, but the expanding Crab nebula showed doubled lines in VM Slipher's spectograph

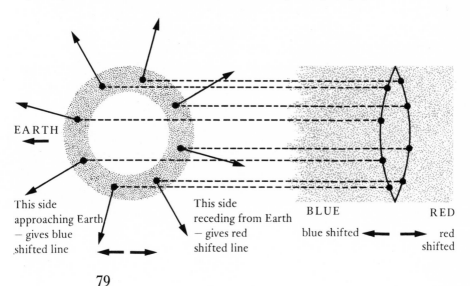

EARTH

This side approaching Earth – gives blue shifted line

This side receding from Earth – gives red shifted line

BLUE   RED

blue shifted ← → red shifted

79

In 1913 V. M. Slipher turned his spectrograph towards the Crab Nebula and photographed its spectrum. Whereas visual observers, using low-powered visual spectroscopes, had seen individual spectral lines, on his photographic plate he was able to distinguish that every spectral line was doubled: it appeared twice, once shifted a small amount towards the blue end of the spectrum and once to the red. He immediately recognized that this was a manifestation of the Doppler effect and that it meant that the light from the Crab Nebula came from two parts, one of which was receding from us and one of which was moving towards us. He correctly deduced that this was because the Crab Nebula

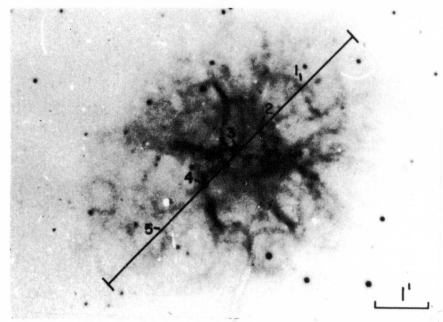

This extraordinary photograph is of the far side of the Crab Nebula. It was taken by Royal Greenwich Observatory astronomers D. McMullan, P. Wehinger and R. Fosbury using a colored filter which selected only light from the redder Doppler-shifted part of the nebulium spectral line from the receding side of the Crab. Light from the approaching side whose nebulium line is Doppler-shifted to the blue was excluded.

80

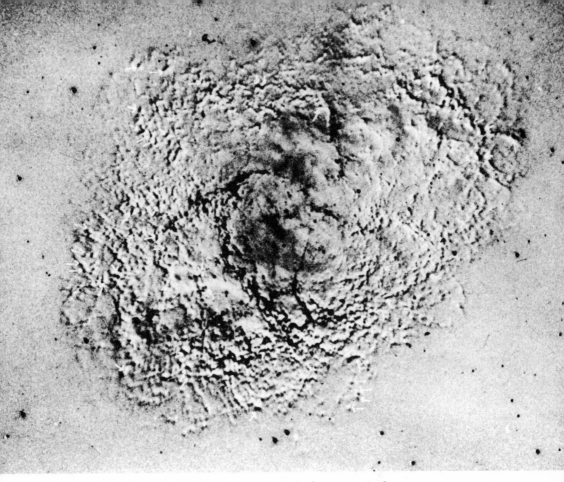

*The expansion of the Crab Nebula over a period of 14 years is shown
graphically in this photograph made by Virginia Trimble. This is a double
exposure of a pair of prints taken with the 200-inch Mt. Palomar telescope in
1950 and 1964. The stars scattered over the photographs were superimposed
precisely so any movement in the intervening years would be clearly seen.
Trimble was thus able to confirm that the filaments were close together at a
central point some 800 or 900 years ago.*

was expanding, with the nearer side approaching us and the farther side
going away, at speeds up to 1000 km/s.

Slipher's observations of the Crab Nebula's spectrum were repeated
by R. F. Sanford at the Mount Wilson Observatory in 1919. Then in
1921, C. O. Lampland photographed the Crab Nebula with the 40-inch
Lowell reflector and compared that photograph with an earlier one. He

81

was able to see that changes had taken place in the Crab Nebula. This provoked speculation that the changes in appearance were a consequence of the high expansion speeds but Lampland could not distinguish between changes in brightness in individual parts of the nebula and motions caused by expansion, because his pictures had been taken on too small a scale.

Then John Duncan photographed the nebula with the 60-inch reflector at Mount Wilson and compared his photograph with an excellent one made in 1919 by G. W. Ritchey with the same telescope. When Duncan compared the two photographs, he was able to see the changes that Lampland had announced, and he was able to tell from his larger scale photographs that many of the outer filaments and parts of the nebula had unmistakably moved outwards. Thus the nebula was actually seen to be expanding, and the spectrographic results were clearly confirmed.

Duncan published his result in 1921, and in the same year, by a remarkable coincidence, the Swedish astronomer K. Lundmark published a list of novae that had been observed by the Chinese. Number 36 in his list was the supernova of 1054. Lundmark noted that the position of that supernova was very near to M1, the Crab Nebula.

Identifying the Crab

None of the other astronomers who had investigated the Crab Nebula made the necessary mental jump of connecting the supernova of 1054 with the expansion of the nebula. This was left to one of the most famous names in astronomy, Edwin Hubble, in 1928. Hubble's name is usually associated with objects outside the Galaxy—he first measured the expansion of the Universe, and the "Hubble Time" defines the age of the Universe itself. He was one of the leading observers of his day, and investigated many nearer astronomical objects, including the Crab. By comparing the size of the Crab with its observed rate of expansion, Hubble was able to estimate that some eight or nine hundred years had elapsed since the expansion began.

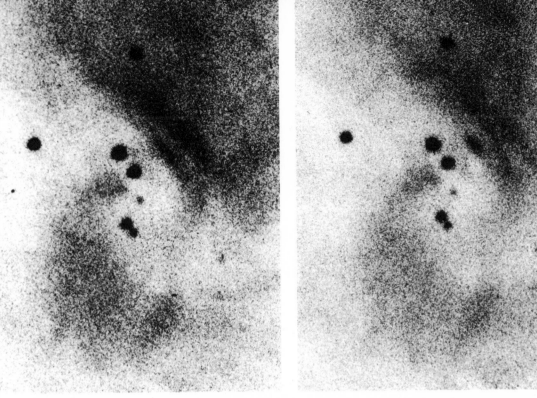

Photo (left) of the central area of the Crab Nebula was taken about the time of the abrupt period decrease of the Crab pulsar in September 1968. By the time photo (right) was exposed, 124 days later, the Thin Wisp had moved out towards the wisps and Wisp 1 had become separated from the others. These Lick Observatory photos (shown as negatives) taken by J. Scargle confirmed Baade's earlier suspicion that "waves of activity" spread into Crab Nebula from the stellar remains of the supernova of 1054, and linked their cause to the Crab pulsar spin-ups.

Hubble was the first to point out the coincidence of position *and age* between the Crab Nebula and the supernova of 1054. He published his inspired guess in a series of popular essays on astronomy and it escaped the attention of professional astronomers. It was not until two Dutch scholars, one an astronomer and one an orientalist, worked on the problem during World War II, that the identification became accepted.

The astronomer was Jan Oort, a leading contributor to knowledge about our Galaxy, and the orientalist was J. J. L. Duyvendak. Working in Holland under the German occupation, they sent their discoveries via Sweden, a neutral country, to N. U. Mayall in the United States.

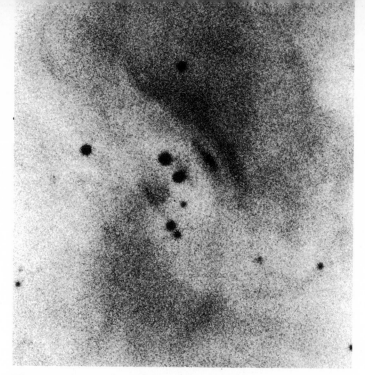

The central area of the Crab Nebula is shown in this photo taken in 1968 by J. Scargle with the Lick 120-inch telescope. In this negative print, brighter objects appear blacker. The pulsar is the lower right (southwest) of the two central bright stars, touched by the Thin Wisp. The bright nebulous area to the upper right is subdivided into two of the three Wisps (the third was not visible at the time this photo was taken). Directly opposite the area of the Wisps is the Anvil, also subdivided into two wisps on this occasion.

Mayall published the Chinese descriptions of the supernova, which were discovered by Duyvendak, in an astronomical discussion in 1942—a remarkable example of the way in which science transcends political divisions, even during wartime. A few years previously, at the Lick Observatory, Mayall had obtained excellent spectra of the Crab Nebula, showing very clearly its expansion as evidenced by the Doppler effect. He found that the speed of expansion was 1300 kilometers every second. Expanding at this speed for eight or nine hundred years, the Crab Nebula had become about seven light years in diameter.

Knowing the true size of the nebula, astronomers could then calculate just how far away it is. Since this method is one often used by astronomers to determine distance, it is worth explaining exactly how it operates.

Consider some object with a fixed size, say a ball 1 foot in diameter. At a distance of, say, 10 feet the ball has a certain apparent size. At twice the distance its apparent size has diminished to half. It appears smaller because of its distance although we know that it still is 1 foot in diameter. Clearly, then, there is a relationship between the distance of an astronomical object, its apparent size and its real size, and if astronomers know any two of these quantities they can calculate the third. In the case of the Crab Nebula, Mayall determined its true size by multiplying the

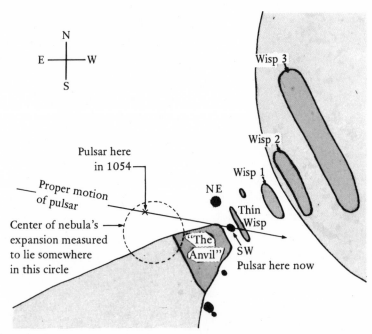

Waves of activity spread from the southwest star SW into the Thin Wisp, The Anvil and the three Wisps 1–3. The SW star was at X in 1054, within the circle which contains the center of the Crab Nebula's expansion. These two facts pointed out the SW star as the active star in the Crab Nebula before it was determined to be the pulsar.

expansion speed by its age, and measured its apparent size in angular degrees, and so could find its distance.

The best modern determination of the distance of the Crab Nebula is 5500 light years, meaning that light takes that length of time to travel from the nebula to Earth. The supernova which the Chinese observed in 1054 had actually exploded in around 4500 BC, and the light from the explosion had been traveling for about 5500 years to reach the Earth on July 4, 1054.

Measurements of the expansion of the Crab Nebula tell astronomers not only when the expansion began, but also the place where it began. The filaments had exploded outward and the place from which they radiated is the place where the explosion occurred. The center of the explosion lies near two stars at the middle of the nebula.

Because of their positions in the sky they became known as the southwest star and the northeast star. The expansion center is actually now equidistant from both stars. However, the southwest star has a so-called *proper motion* of its own. (The word "proper" is not used in any sense of moral rectitude, but because the motion belongs to the star itself and has nothing to do with the motion of the Earth, as, say, does its rising and setting.) The northeast star has no discernible proper motion. In their 1942 paper on the Crab Nebula, Walter Baade and Rudolph Minkowski took into account the motion of the southwest star since 1054 and found that *this* star was closer to the original site of the explosion. The northeast star is probably not in the nebula at all, but lies in that direction by chance.

The southwest star lies in a hole in the center of the Crab Nebula, surrounded by little wisps, bays and filaments which seem to change in brightness and position. The changes were among those first seen by C. O. Lampland in 1921. Walter Baade studied the changes in the center in detail and found in 1945 that the brightness changes gave the impression that light was rippling in waves outwards from the southwest star. He took photographs with the 200-inch telescope to prove his point, but had not completed analyzing them when he died in 1960. Guido

Münch and Jeffrey Scargle subsequently used these plates and later ones they took themselves to confirm Baade's work.

Baade and Minkowski strongly suspected in their 1942 paper that the southwest star was not only the central remnant of the supernova explosion, but that it was still affecting the central regions of the nebula. They were correct. The southwest star is in fact the Crab Nebular pulsar, as we shall see in the next chapter, and waves of activity propagate from this star throughout the nebula.

The accelerating Crab

The most accurate estimates of the Crab's expansion actually put the date at which it began to within 10 years of AD 1140, significantly different from 1054. Has there been some strange mistake? In fact, astronomers interpret this as evidence for another of the Crab Nebula's peculiarities, and one which once fooled Baade himself.

The estimate of the date when the expansion began is based on the assumption that the nebula has been expanding throughout its history just exactly as fast as over the last 30 years (the time interval between the photographs used to measure the expansion speed). The identification of the Crab with the 1054 supernova is so strong that this assumption has to be wrong. But, rather surprisingly, the nebula is not slowing down: it is speeding up. If it were slowing its present expansion speed would now be smaller than in the past and we would mistakenly say that the explosion occurred earlier than 1054. Since it appears from the expansion rate that the explosion occurred a hundred years later than 1054, the present speeds are larger on average than the speeds have been over the last nine hundred years.

This was first discovered by Walter Baade, using Duncan's measurements, but at first he thought that the observations were not accurate enough. He simply could not believe the result because he would have expected the expansion to slow down as the nebula crashed into the surrounding interstellar material. But it does seem to be the case that

87

the nebula is accelerating. What that implies is that some extra energy is available from somewhere to drive the expansion at faster and faster speeds. The initial energy of the supernova explosion has caused it to start expanding but some additional energy has been pumped into the nebula to make the expansion faster.

Before the mystery of the source of the energy accelerating the filaments could be solved, radio astronomers found further proof that there was an active powerhouse in the Crab, and that it wasn't just using up its inheritance of energy from the supernova explosion itself.

The Crab among the radio stars

Astronomy, and particularly radio astronomy, is full of unexpected discoveries, which pop up through serendipity, as the result of looking for something else. Indeed, the man whose name is now indelibly linked with beginning radio astronomy did not set out to do anything of the sort. The man was an engineer at the Bell Telephone Laboratories, and his name was Karl Jansky.

Bell were interested in the hiss which represented the ultimate limit to the sensitivity of radio reception and transmission, and therefore set Jansky to investigate atmospheric interference. To do this, he built an antenna which he called the Merry-go-Round, because it could be rotated to track down the source of the hiss. In December 1931, Jansky noticed a source of radio interference whose intensity cycled with a period of 23 hours 56 minutes, that is, it was keeping pace with the rotation of the Earth with respect to the stars and not recurring every 24 hours which is the period of rotation of the Earth with respect to the Sun. This led Jansky to conclude that the hiss was of cosmic origin. It peaked in the Milky Way constellation Sagittarius.

Because of commercial pressure, Jansky was unable to continue this investigation, and the first radio maps of the Milky Way were made by an American amateur astronomer, Grote Reber, in the early 1940s, using a home-built backyard antenna. They showed a broad swathe of radio

88

noise coming from the Milky Way, most strongly towards the galactic center in Sagittarius. There were two subsidiary peaks in the constellations Cygnus and Cassiopeia blending into the general Milky Way.

Radio astronomy had been studied during World War II largely because natural cosmic radio noise affected radio reception and the operation of radar. With the ending of the war, it became possible to investigate these phenomena more fully and the opportunity to do so was provided by large quantities of surplus wartime radio and radar equipment. A radio telescope, after all, is no more than a sensitive radio receiver coupled to a directional antenna, such as those used for radar.

A team in England led by John Hey had found during wartime research that the Sun was a source of radio noise when there were large sunspots, and that meteors showed on radar. After the war this team mapped the intensity of radio waves along the Milky Way and noticed

A radio telescope's view of the Crab. The basic map has brightness contours only, which here have been shaded in.

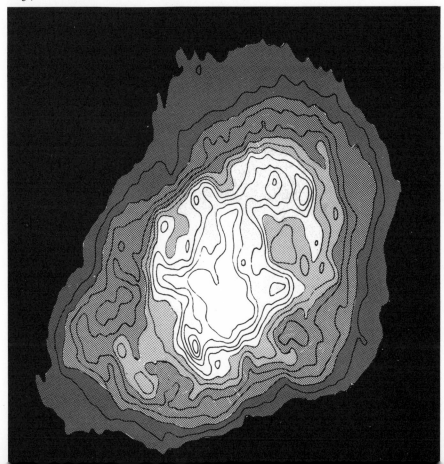

the bright pointlike source of radio waves in Cygnus. Hey's radio telescope had a narrower beam than Reber's so that the Cygnus source stood out as more pointlike, but the blurriness of Hey's telescope's beam was still 16 times the area of the Moon: its resolution was 2°.

Imagine the optical appearance of the night sky seen with a 2° beam. In fact you can simulate such a beam by looking through very defocused binoculars at the sky. Seen like this, the Moon has no surface detail at all. Its phases cannot be distinguished; all that can be seen is its change of brightness through the month. The appearance of the Milky Way is about the same whether seen through these binoculars or not. The blurred shapes of all but the brightest stars merge into the Milky Way and are indistinguishable from it. The brightest stars, however, can be seen as blurred disks. The view hints at crisper detail beyond the resolution of the binoculars.

The telescope that looked to sea

Arguing by such an analogy, some radio astronomers put effort into determining whether Hey's radio source in Cygnus was truly pointlike, and whether there were other radio sources like it. But how could they make a radio telescope that pointed well enough? In Australia, John Bolton and G. J. Stanley used a radio telescope situated on Dover Heights, an eastern suburb of Sydney. They set the telescope on a cliff top overlooking the Pacific Ocean and directed it towards the radio stars as they rose. It received radio radiation not only directly from any radio stars that rose above the horizon, but also from their reflection in the sea. In effect, it functioned in some ways like two connected radio telescopes, the real one 300 feet above sea level on the cliff top, and another one, 300 feet below the sea. Such a radio telescope, called an interferometer, has the same ability to discriminate fine detail as a single vast radio telescope 600 feet in diameter.

With this telescope, Bolton confirmed the existence of the Cygnus radio source and also saw a radio source in Taurus, which he named

Taurus A. His method of determining a radio source's position in the sky consisted in looking at the time at which it rose above the horizon and at the rate at which it ascended above it. However, there was a problem analogous to atmospheric refraction. The effect of air on the light waves from stars at the horizon is to make them visible before they have actually risen above the horizontal. This effect is seen in an exaggerated form in a mirage when light from an oasis below the horizon can be refracted into the gaze of a desert traveler. Normally, atmospheric refraction causes stars and the Sun to rise at least two minutes early and set at least two minutes late. Another effect is that changes in the degree of refraction cause stars near the horizon to twinkle more than stars overhead. Radio refraction in the ionosphere above Earth is similarly highly variable and is more severe than optical refraction in the atmosphere. Moreover, when radio refraction is strongest, it is least variable and causes least twinkling in the appearance of the radio stars.

Bolton selected the clearest records of Taurus A rising, and in doing so he had selected those with abnormally large radio refraction. Consequently, the first estimate of the position of Taurus A was a long way in error. Compensating for this refraction in later experiments, Bolton took his interferometer to New Zealand, where he observed Taurus A rise from the eastern side of the island, and set from the western side. Any error in time of star-rise was compensated by an equal and opposite error in the time of star-set. Bolton then looked to see what optical sources were in that part of the sky, using his only suitable reference work, a star atlas used predominantly by amateur astronomers, *Norton's Star Atlas*. There he found marked M1, the Crab Nebula. Thus in 1948, a radio star was identified with a visible object for the first time.

Bolton and Stanley later used a genuine twin telescope interferometer to observe Taurus A when it was well above the horizon and the effects of refraction were small. They tied the position of Taurus A down to within 15 arc minutes, an area one quarter of that of the Full Moon.

A question immediately arose: was the radio emission coming from the Crab Nebula itself or from a star embedded in the nebula? A two-pronged attack brought the first answers. The initial approach was by various technical advances in making radio interferometers of finer and finer resolution, by placing the radio antennae farther and farther apart. A Manchester group led by Robert Hanbury Brown worked with separations up to 4 km. In Australia, Bernard Mills used two radio telescopes up to 10 km apart to obtain a viewing beam about 1 arc minute wide. The second line of attack was afforded by a lucky chance.

Taurus is not only a Milky Way constellation, it is also on the ecliptic, the yearly path of the Sun, Moon and planets. The Moon, in fact, eclipses the Crab Nebula, and it is possible to see the gradual fading of Taurus A as the Moon moves in front of it. Eclipses in 1956 showed that the radio emission comes from the whole of the nebula, because it is covered gradually as the Moon advances across it. With the technically superior interferometers now available at Cambridge, A. S. Wilson has made the most detailed maps of the radio emissions from the Crab, showing that its shape is remarkably like that of the white light component of the Crab Nebula, with the same broad hump, dissected by the bays and valleys that show at optical wavelengths.

X-rays from the Crab

In 1963, the Crab Nebula was discovered to be emitting X-rays. X-radiation is a form of radiated energy, like light, but a thousand times more energetic. When X-rays shine from outside the Earth onto atoms in the upper atmosphere, such as oxygen and nitrogen, the X-rays are quickly and readily absorbed. In this process, electrons orbiting deep within the atoms take up the X-ray energy, causing the electrons to be ejected from the atoms. X-rays traveling at sea level are typically completely absorbed by air after a few hundred feet or so. (Similar X-rays are used in X-ray machines to photograph bones of the body: the areas where X-rays have been absorbed, by just a few inches of flesh and bone, show up as shadows on a photograph.)

Because X-rays are so readily absorbed, the only way astronomers can detect cosmic X-ray sources is by flying X-ray detectors above most of the atmosphere. Balloons can carry X-ray telescopes above a good deal of the atmosphere, but by its nature, a balloon, needing air to support it, cannot escape completely above the air. Only the most penetrative and energetic X-rays can be detected in this way. (Balloons also have the astronomical disadvantage that they obscure the most interesting and least absorbed part of the sky, immediately overhead.)

Rockets have the ability to carry instruments completely above the atmosphere in long parabolic trajectories, without going into orbit, and it was by this means that the Crab Nebula X-ray source was detected in 1963. A rocket flight, however, lasts for just a few minutes, as long perhaps as a total solar eclipse, and costs so much that it occurs about as rarely.

Satellites, launched into orbit around the Earth above the atmosphere, last for several years, so that although they are more expensive to send up, they are far less expensive per minute of observing time than rockets. However, the first satellite launched exclusively for X-ray astronomy, *Uhuru*, was not launched until 1971.

The X-ray telescope with which a group of X-ray astronomers at the Naval Research Laboratory, Washington, D.C., discovered the Crab Nebula in 1963, accepted X-rays from an area of the sky 20° wide. The year before, a group at the American Science and Engineering company had flown an experiment which detected an X-ray emitting region in the Milky Way in Taurus, and the NRL rocket was commanded to sweep over this area. As the NRL X-ray telescope slewed over the Crab Nebula, it recorded a maximum in the number of X-ray counts, and the scientists found that the peak was somewhere within 2° of the Crab. In this way, the Crab became the first X-ray source to be identified with a known celestial object (excepting the Sun), just as it had been the first identified cosmic radio source.

By 1969, the Crab X-ray source had been the subject of more than 30 rocket and balloon flights carrying X-ray equipment. It is interesting to estimate the cost of the effort put into studying the Crab's X-ray photons, assuming that an X-ray experiment costs something on the order of $100,000 and collects perhaps 10,000 to 100,000 X-rays from the Crab.

The Moon pinpoints X-rays

The NRL group followed up its work with a further rocket, launched so that it would be looking at the Crab during its eclipse by the Moon on July 7, 1964. Three minutes after launch, the Crab was seen beginning to disappear as the Moon passed in front of it, and it slowly faded away to nothing over the next three minutes. During those three minutes, the edge of the Moon had scythed across two arc minutes of the sky, so that the NRL group had shown that the X-ray emitting region was approximately the same size as the optical and radio Crab Nebula.

The "seasons" during which the Moon eclipses the Crab recur every 10 years. A group of X-ray astronomers at the Lawrence Livermore Laboratory of the University of California observed the next lunar eclipse of the Crab in 1974. They were able to tell that the lower-energy X-rays in fact came from an area somewhat smaller than the optical nebula, and centered somewhat west of the star suspected as the center of the Crab, and near the brightest of the wisps, which Baade and later

Following a spiral path around a line of magnetic force, an electron in the Crab Nebula radiates radio, optical and X-radiation by the synchrotron process.

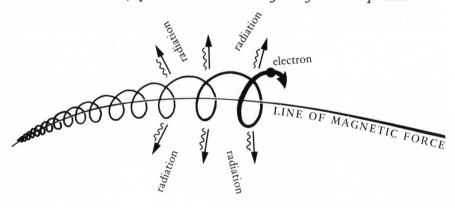

94

Scargle had found to shift position and change in brightness. Another experiment, this time aboard a balloon operated from MIT, showed the very high energy X-rays to come from an even more compact area centered on the wisps. The X-radiation therefore has a somewhat more intimate connection with Baade's southwest star and its immediate surroundings than optical and radio radiation, but if unearthly beings with strange eyes were to study only the radio or X-ray Crab Nebula, they would picture it in much the same way that we do: the invisible Crab mimics the visible one.

What makes the Crab shine?

The Crab Nebula looks similar in light, radio waves and X-radiation, unlike most other nebulae. And, indeed, all the radio, X-ray and the optical emissions are caused by the same phenomenon, known as synchrotron radiation. When the Russian astrophysicist Iosif Shklovsky proposed this explanation in 1953, the theory was at first thought to be absurd. The model it proposed was of a Crab Nebula full of fast electrons gyrating about strong magnetic fields and radiating their energy over a wide spectrum stretching from radio through optical. Shklovsky predicted that the radiation would be polarized so that when the optical radiation was photographed through a polarizing filter, the nebula would appear streaky along the lines of the magnetic forces which pervade it. This was found to be the case, and the theory of synchrotron radiation was vindicated.

Now, a characteristic of electrons radiating by the synchrotron mechanism is that they lose their energy relatively quickly. The lifetime of a fast moving electron in the Crab Nebula (that is, the time in which it radiates half its energy) is much less than the 900 year age of the nebula. Therefore, the fast electrons cannot have been spiraling about the nebula since the original supernova explosion: they must have been injected into the nebula since then. The existence of the synchrotron radiation demands that energy must be pumping into the nebula in the

95

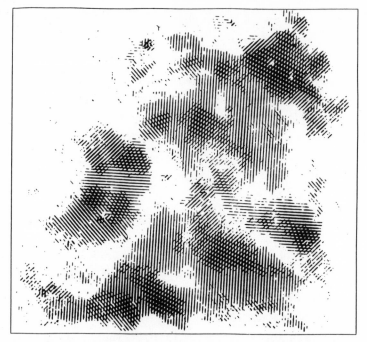

*Fritz Zwicky described the polarization pattern of the Crab Nebula as a
"basket weave." It is well shown in this photographic representation by
F. Gieseking of the Bonn University Observatory. Polarized light comes from
electrons gyrating in the Crab's magnetic field which runs perpendicular to the
lines on this diagram.*

form of fast electrons at this moment. Remember that, from the rate of
the Crab's expansion, there is other evidence that some powerhouse is
still functioning in the Crab. Until the discovery of pulsars in 1968 the
source of this extra energy was a mystery.

VI *Discovering pulsars: heartbeats of supernova remnants*

The story of the discovery of pulsars in 1967 is a classic among the many scientific tales of perspiration, inspiration, and just plain luck. Pulsars linked the practical world of the observing radio astronomer with that of the theoretician, who for years had been talking about mysterious objects called "neutron stars." And, in explaining how they pulsed and how they emitted radio waves, astronomers found that they naturally provided an explanation for the amazing expansion and acceleration of the Crab Nebula.

The discovery had the added drama that, for a time, it seemed to be evidence for extraterrestrial life, although by the time the news of the first pulsar was published, its discoverers were already quite sure that the signals were not artificial.

Even the instrument with which pulsars were first detected has curiosity value in its own right. While most radio telescopes are measured by their size in feet, meters, or even miles, this one has a splendidly archaic name, which never fails to confuse foreign astronomers: it is the $4\frac{1}{2}$ acre telescope at Cambridge, England.

It was designed for a purpose which had nothing to do with the regions of space where pulsars lurk. It was, and in its present enlarged form still is, designed to pick out those enigmatic radio sources known as quasars, by means of their scintillation or twinkling.

97

Every amateur astronomer knows that you can easily tell a planet from a star because a star, being a point of light, usually twinkles whereas a planet doesn't. Twinkling is caused by unsteadiness in the Earth's atmosphere. A planet twinkles much less, if at all, because in a telescope it has a discernible disk, even though to the eye it may seem the same size as a star.

Much the same sort of thing happens with radio sources, though radio astronomers use the word "scintillation" instead of "twinkling." The medium causing the disturbance is thin gas called plasma. Not only is plasma found in the atmosphere of the Earth but also far out in the solar system, between the planets.

For Antony Hewish and his team, who first picked up interplanetary scintillation from pointlike radio sources in 1964, its importance was that it enabled the tiny, distant quasars (quasi-stellar radio sources) to be picked out from nearer, apparently larger, sources.

To get the sensitivity necessary to distinguish rapid fluctuations of signal, Hewish needed a radio telescope with a large collecting area. This he achieved by simply setting up wires on poles, covering a paddock of $4\frac{1}{2}$ acres. This had the required ability to pull in faint signals, but it lacked the direction-finding discrimination of a dish-type antenna. By July 1967, the new telescope was ready to begin recording. It was designed to scan a large part of the sky in one week. The equipment was built to emphasize the scintillation rather than, as was usual, to de-emphasize it, and it was able to respond in as short a time as one tenth of a second to fluctuations in the brightness of a radio source. Hewish wanted all the radio sources which it found to be plotted on a map of the sky, so that terrestrial manmade interference, which would appear randomly, could be sorted out from the truly extraterrestrial twinkling radio sources which would recur at the same celestial coordinates. The person to whom he assigned the job of analyzing the data from the instrument was a graduate student, Jocelyn Bell.

Scruffy little green men?

Analyzing the data from the new telescope was no small task. The instrument produced 400 feet of tape from each scan of the sky, 100 feet every day. Bell's job was to examine every signal, discarding such manmade phenomena as aircraft transmissions and foreign television stations and mapping out the true extraterrestrial signals. By October, she was 1000 feet behind current chart production and yet, fortunately, she did not relax her standard of attentiveness.

It was in October that Bell noticed what she called "a bit of scruff." It was passing through the beam near midnight when the interplanetary scintillation normally falls to a low level, because at this time the radio telescope is pointing to the outer edge of the solar system where the plasma is least dense. Bell's account says, "Sometimes within the record there were signals that I could not quite classify. They weren't either twinkling or manmade interference. I began to remember that I had seen this particular bit of scruff before, and from the same part of the sky."

The source seemed to be recurring every 23 hours 56 minutes, and only objects fixed among the stars recur every 23 hours 56 minutes. Manmade interference, though, tends to recur on a 24 hour schedule, because daily life is ordered by the Sun. The moment when Bell recognized that the bit of "scruff" was more than a single piece of interference, but had actually occurred before at the same celestial coordinates, proved to be very important. Bell describes her reaction: "When it clicked that I had seen it before I did a double take. I remembered that I had seen it from the same part of the sky before."

Looking back at the records, Bell was able to prove that she had, in fact, seen it two months before, in August. She then discussed the signals with Hewish. They decided to use the observatory's fast recorder to get a clearer picture of the nature of the signals. When the fast recorder became available in mid November, Bell was given the job of trying to catch the signals and record them. For some days she was unsuccessful.

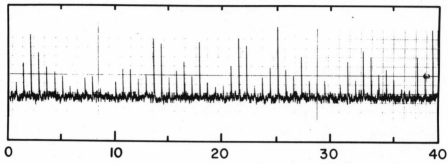

How radio astronomers see a pulsar. This is a recording of the signals from radio pulsar 0329 + 54, observed with the National Radio Observatory's 92-meter transit telescope by Richard Manchester. Each spike on the chart is a pulsar flash recurring at 0.714 second intervals.

At this point, Hewish thought that the signals were from a randomly occurring flare star and that it was unlikely that they would see it again. Bell persevered, however, and at last managed to catch a satisfactory recording which showed clearly that the "scruff" was a burst of pulses almost exactly $1\frac{1}{3}$ seconds apart, similar to many kinds of terrestrial interference. When she telephoned to Hewish to tell him what she had found, he said, "Oh, that settles it. It must be manmade."

Nonetheless, Bell and Hewish continued to make recordings of the "scruff." The main problem was still that the pulses were keeping sidereal time, recurring with the 23 hour 56 minute period. Were the bursts truly sidereal, or were they being made artificially with a sidereal period? The only people on Earth who could conceivably be imitating sidereal time would be astronomers, though no one could guess why they would want to make bursts of pulses like this. Inquiries at other observatories failed to reveal any program which could account for the signals. Searching around for sidereal explanations for the pulsing signal, Hewish and Bell considered whether known variable stars could cause it. The trouble was that the fastest variable star known had a period of about a third of a day. How could a star throb with a period of 1.337 seconds?

Caught in the dilemma that the pulses were extraterrestrial, but seemed to be artificial, the Cambridge astronomers began to consider a new possibility: were the pulses being manufactured in space by an extraterrestrial civilization? By mid December, they had proved that the pulses recurred very regularly indeed, staying on schedule to one millionth of a second. In a half-joking way, Bell's colleagues began to refer to the star as LGM-I. But why would Little Green Men manufacture and broadcast repetitive signals like this? Most radio signals change, in order to convey information; the constant ones are navigational aids like the LORAN signals (for LOng RAnge Navigation). Was LGM-I an interstellar navigational beacon?

If the pulses were being manufactured artificially by an intelligent civilization, the manufacturers presumably lived on a planet. If the signals were coming from a planet, they would show the effects of a *Doppler shift*. The Doppler shift causes a bunching effect of repetitive signals as the transmitting object moves towards the recipient, and a spacing out effect as it moves away. The Cambridge astronomers had, in fact, already observed a small change in the timing of the signals, caused by the motion of the Earth around our Sun. Could they see the equivalent effect, at a different period, caused by the transmitter itself being on another planet orbiting its own sun? Bell recorded in her log, "We are working on the Doppler shift of the pulses to see whether the source is stationary or moving round a sun. There is no 4C (Cambridge catalog) source with the same coordinates, nor any other source that we know of." In the event, no Doppler shift was seen, other than that caused by Earth's motion. The radio source was therefore not on another planet. The little green men became less likely.

The Cambridge radio astronomers used the amount of Doppler shift on the period of the pulses to estimate the source's position in the sky because, as we have noted, the $4\frac{1}{2}$ acre telescope lacked direction-finding precision. Suppose that a pulsar with a period of one second lies in the plane of the Earth's orbit. On a certain day of the year, the Earth will be traveling directly towards it with its orbital speed of 30 kilometers per second. This will cause a decrease of the period by one ten thousandth of a second to 0.9999 seconds. Six months later, the Earth on the other side of its orbit will be traveling away from the pulsar at 30 kilometers per second, and the period will be lengthened to 1.0001 seconds.

On the other hand, if this imaginary pulsar is in a direction perpendicular to the Earth's orbit (a direction known as the pole of the ecliptic), the Earth will throughout the year be traveling across the line of sight to the pulsar, no Doppler shift will be observed and the period will always be 1.0000 seconds.

Such small changes may seem impossible to measure, particularly when we see the indistinct signal from a weak pulsar. But we are able to take the average of hundreds or thousands of pulses. After 10,000 pulses of the hypothetical pulsar, each one delayed by 0.0001 seconds, the combined delay will be one whole second. Ten thousand seconds is not quite three hours, so in a brief space of time it is possible to measure the rate of a pulsar to much better than 1 part in 10,000. From the size of the Doppler shift of the first pulsar, the Cambridge team estimated its position and that it was unmoving, to an accuracy of about two arc minutes. Just as Brahe had shown for the supernova of 1572, and Kepler for the supernova of 1604, and with the same accuracy, the Cambridge astronomers had demonstrated that the pulsar had no discernible parallax and must be farther than the edge of the solar system.

Measuring the pulsar's distance

The Cambridge team then proceeded to make some measurements which were possible with two radio receivers operating simultaneously, but tuned to different radio frequencies.* Pulses

* The frequency and wavelength of radio signals—or any other kind of waves—are linked. The frequency is the number of waves passing a point in a given time, while the wavelength is the crest-to-crest distance of the waves. As the wavelength gets shorter, so the waves get more frequent and the frequency is higher.

observed on the two frequencies arrived at different times, a pulse traveling on the longer wavelength radio arriving later than the same pulse traveling on the shorter wavelengths. The delay between the two signals showed that the radio frequencies traveled through interstellar space at different speeds, a phenomenon called *dispersion*.

If interstellar space were truly empty, all radio waves would travel at the speed of light. But there are free electrons in interstellar space, produced from the ionization by starlight of atoms of interstellar gas, such as sodium. Starlight passing near a sodium atom can interact with the atom, causing it to eject the loosest of its electrons. Interstellar space is thus not empty: it contains a plasma, a low density gas containing unattached electrons, and radio waves traveling through it, especially low wavelength ones, are slowed from the speed of light by tiny amounts.

The amount by which the radio waves are slowed down depends on the density of electrons in the plasma. In interstellar space, radio waves are slowed down by typically one inch per second from the 186,000 miles per second speed of light. The time delay which this causes between pulses observed at different radio wavelengths is called the *dispersion measure*, and depends on the square of the electron density multiplied by the distance to the star. By assuming a value for the electron density in space (it had previously been studied by other means), the astronomers were able to estimate the distance of the pulsating radio source to be 200 light years, placing it among the stars rather than close to the solar system or outside the Galaxy altogether. Though the slowdown of radio waves by the interstellar plasma is a small decrease from the speed of light, the distance that the waves travel is so large that the difference in speed causes a measurable delay of typically one second.

How big is it?

At this time the radio astronomers were also able to make an estimate of the size of the pulsating star (it was this latter phrase which was contracted to "pulsar"). They measured the length of the individual pulses at a particular radio wavelength and found that each pulse lasted for about 16 milliseconds, only two per cent of the period between the pulses. Each cycle of the pulsar was a brief flash with a relatively long time between flashes, and whatever was causing the flash had to emit the light all within a time of 16 thousandths of a second.

The duration of the flash arose from two causes. Consider two parts of the flashing region: let us suppose that each part simultaneously makes a brief mini-flash, but let us suppose that the two parts are separated on our line of sight by a distance, say d. At the speed of light, c, the rear flash takes an extra time d/c to reach Earth. Therefore when the flashes are observed at Earth they appear as a pair of pulses separated by an interval equal to the time taken by light to travel the distance between the two parts of the flashing region.

Clearly, if the flashing region has many components spread over a distance along the line of sight, the flash observed would be smeared out over the time that would be taken by light to travel that distance. If the flashes from the component parts are not really simultaneous then the spread of the pulse will be even more than the light-time along the depth of the source. Therefore, the duration of the flash of the pulsar as observed on Earth shows the maximum extent of the pulsar's depth. The pulsar observed by Bell and Hewish must be smaller than 16 light milliseconds in depth: less than 3500 miles.

What sort of object could emit rapid, energetic radio pulses, yet be smaller than the Earth? This is much smaller than any ordinary star, but it is about the size of certain very condensed stars. For many years, astronomers had been aware that there are some very strange stars indeed in the sky. Among the strangest are the white dwarfs, the end products of stars in which the normal energy processes have ended, leaving them at the mercy of their own gravity. These stars collapse upon themselves, forcing their atoms into a super-dense state called *degeneracy*, and resulting in the entire mass of the star being packed into a body little larger than the Earth. White dwarfs are not uncommon in the sky—one is a companion of the bright star Sirius, for example. But theoretical astronomers did not rest content with white dwarfs. There should be even stranger stars, they said, smaller and denser than white dwarfs. They called them neutron stars. For 30 years, neutron stars were the theoretical solution to a problem which did not exist: mathematically it had been shown that they could exist but no trace had been found of a real example. As evidence grew that the newly discovered pulsar was a natural phenomenon, the thoughts of the Cambridge astronomers turned to stars of this kind. But there were still more discoveries to be made.

More pulsars turn up

In December, Bell discovered a second pulsar. "I was working in the evening analyzing charts. I saw something which looked remarkably like the bits of scruff we had been working with. This was in a bit of sky which wasn't very easy for the telescope to look at, but there was enough to confirm that there had been scruff.

"That particular bit of sky was due to go through the beam at one in the morning. It was a very cold night and the telescope doesn't perform very well in cold weather. I breathed hot air on it, I kicked and swore at it, and I got it to work for just five minutes. It was the right five minutes, and at the right setting. The source gave a train of pulses but with a different period of about one and a quarter seconds."

Finding a second pulsar made it even less likely that the transmissions were artificially produced by another civilization. "There wouldn't be two lots signaling us at different frequencies. So obviously we were dealing with some sort of very rapid star. I threw up another two sometime in January." "Throwing up" another two had involved searching back through the three miles of tape which had accumulated. Now that they had this further evidence that the signals were from a natural galactic source, the astronomers felt ready to publish the news of the discovery of the first pulsar. They wrote the paper describing the observations, and submitted it to the science journal *Nature* eight weeks after the recognition that the first radio source was pulsating. Shortly afterwards, on February 24, 1968, the paper on the first pulsar was published in *Nature*, authored by Hewish, Bell, and their colleagues J. Pilkington, P. Scott and P. Collins.

The story of the discovery of pulsars has a pattern which recurs regularly in the history of science, though not as repetitively as pulsars themselves. Built for a completely different purpose, Hewish's radio telescope picked out the pulsars by chance. Bell noticed the signal from the first pulsar because, though it resembled both spurious interference and the normal twinkling of radio stars, its characteristics did not quite fit with either because it recurred in the same part of the sky and it twinkled at midnight. Although she was still a graduate student and therefore inexperienced in astronomy, she had a receptive mind and she seems to have been less ready to dismiss the "scruff" as interference than her more experienced colleagues (there is a persistent rumor that pulsars had been seen by a radio astronomer before Bell, but had been dismissed by him). After patiently but quickly bringing out the essential characteristics of the pulsars, the Cambridge group, armed with the confidence that they had discovered further examples, published their data and were able to mention, in their discussion of it, what is now believed to be the kind of star responsible.

A row flares

Like many stories of superb scientific discoveries, this one has parts that could have been scripted by C. P. Snow. The Cambridge group was criticized for sitting on the discovery of the first pulsar for six months and then concealing the discovery of the further pulsars which they had found. Actually there were only two months between the recognition of the repetitive pulses from the first pulsar and submission of their paper to the journal *Nature*.*

* Fred Hoyle's novel *A for Andromeda*, written with John Elliot and published five years before Bell's discovery, contains a remarkable

Professor Sir Bernard Lovell, Director of Britain's Jodrell Bank Observatory said he thought "the Cambridge people had behaved with exemplary scientific discipline in withholding news of their discovery until they were satisfied about the general nature of the objects, but that there was no excuse for a similar delay in withholding information about similar objects which they had discovered." But this incident was of small importance compared with a later row.

In 1974, the Nobel Prize for physics was awarded for the first time to astronomers. It was given jointly to Martin Ryle, Director of the radio observatory at Cambridge for his work in developing new kinds of radio telescopes, and to Antony Hewish for his "decisive role in the discovery of pulsars." The Nobel Prize committees are thought to investigate the circumstances very thoroughly when they make their awards, and they seem to have been satisfied that Bell's part in the discovery did not merit a share in the prize.

In the following March, however, the well-known British astronomer, Fred Hoyle, one-time director of the Institute of Theoretical Astronomy in Cambridge, criticized the way in which the Nobel Prize had been

---

foreshadowing of the discovery of pulsars. A large, new radio telescope picks up from the direction of the constellation Andromeda a "faint single note, broken but always continuing" like Morse code. The radio astronomers deduce from its constant position in galactic coordinates that it is not in orbit in our solar system, nor an artificial satellite. Unlike the real pulsars the signal turns out to be a message, with a lot of "fast detailed stuff" between the dots and dashes. Like the Cambridge radio astronomers, the fictional ones build a fast recorder to record it. A news blackout is imposed by a high-up civil servant, to be broken by the most individualistic of the scientists. The press sensationalize the signal: "SPACEMEN SCARE: IS THIS AN ATTACK?" Perhaps this is another reason why, when pulsars were first discovered at Cambridge, the radio astronomers imposed a news blackout on themselves, avoiding misrepresentation when they were unsure of precisely what they were observing.

awarded to Hewish without including Bell. According to Hoyle, the crucial parts of the discovery were the recognition of the signal as something unusual, and the observation that it was keeping sidereal time. After this, any astronomer would have gone through the same reasoning process and come to the same conclusions as the Cambridge team. Hoyle wrote:

There has been a tendency to misunderstand the magnitude of Miss Bell's achievement, because it sounds so simple, just to search and search through a great mass of records. The achievement came from a willingness to contemplate as a serious possibility a phenomenon that all past experience suggested was impossible. I have to go back in my mind to the discovery of radioactivity by Henri Becquerel for a comparable example of a scientific bolt from the blue.

In a reply Hewish wrote that Bell had been carrying out a program initiated and mapped out by him. He said that her work as a graduate student had been excellent, but that it would be unjust to later graduate students who continued the analysis to suggest that they would not have discovered the pulsar themselves, had they been in her position. Of course, the argument that the next person down the line would have made the discovery anyway if X had not, applies to most scientific work. Very often in great discoveries there is an element of luck: being in the right place at the right time, with the right predecessors. No matter who received the Nobel Prize, Bell actually discovered the first pulsar of the 149 now known.

More discoveries

After the four Cambridge pulsars were found others were discovered by astronomers using radio telescopes at Green Bank, W. Va.; Jodrell Bank; Arecibo, Puerto Rico; and Molonglo in Australia. In a discovery late in 1968 which indicated the connection between pulsars and supernovae, a Sydney University group using the Molonglo radio telescopes discovered a very short period pulsar which lay in the same direction as a source of radio emission called Vela X. Could the two be linked? Vela X had been previously identified by Douglas Milne as the remnant of a supernova which occurred some 10,000 years ago. Milne put Vela X at a distance of about 1700 light years, and the Sydney radio astronomers M. I. Large, A. E. Vaughan and Bernard Mills deduced from its dispersion measure (p. 103) that the pulsar which they discovered was at the same distance. They inferred that the pulsar probably was the stellar remnant of the supernova—the small star left over after the supernova explosion which had ejected the outer parts of the original star into space and made the Vela X radio source. The pulsar had a very short period—only 89 thousandths of a second, and each brief flash of radio waves lasted only 10 thousandths of a second.

Within a few weeks, however, an even shorter period pulsar (still the shortest period known) was found in the center of the Crab Nebula which proved the connection between pulsars and supernovae. This one was discovered in a deliberate search of the Crab for its pulsar by D. H. Staelin and E. C. Reifenstein at the National Radio Astronomy Observatory in Green Bank, W. Va., using a brilliant method which exploited one of the properties that made its detection as a pulsating star difficult—its high dispersion.

As explained previously, pulsar pulses, when observed at different radio wavelengths, arrive at Earth at different times. Now, no radio telescope observes at a single radio wavelength—it always detects radio waves arriving within a small band of wavelengths, called the bandwidth of the radio receiver. (This is true of a domestic radio receiver too: cheaper quality radios receive a wider bandwidth than more finely tuned, expensive ones and may receive the programs broadcast by two radio stations at adjacent wavelengths on the wavelength dial. The two stations may interfere with each other and their individual signals are muddled.)

109

The wider the bandwidth in a radio astronomy receiver, the more radio energy it receives from the sky and as a consequence it can detect fainter radio stars. But, if a radio star is pulsating fast and there is strong dispersion, a pulse from the pulsar can be received at the lower wavelength edge of the bandwidth at the same time as previous pulses are being received at longer wavelengths still in the receiver's bandwidth. As a result the pulses are blurred together and not distinguished or recognized as pulses by the radio telescope.

Staelin and Reifenstein realized that this would occur if they were to search for a pulsar in the Crab Nebula since dispersion is caused by electrons in space and the Crab contains many free electrons. This is evident from its appearance, with red H-alpha filaments and overall synchrotron glow. They decided therefore to use finely tuned narrow bandwidth radio receivers so that they could observe individual pulses, compensating for each receiver's individual lack of sensitivity by coupling them all together in a bank of 50, and looking for a particular pulse as it was received in turn by each of the receivers. It turned out, when they in fact discovered the Crab Nebula pulsar, that it took $1\frac{1}{2}$ seconds for a pulse from the pulsar to sweep through all 50 of their receivers, because of the dispersion. The period of the pulsar turned out to be just 33 thousandths of a second.

The two shortest period pulsars then known had been discovered to be associated with supernova remnants. To this day no other pulsar of longer period has been unambiguously associated with a supernova remnant. Searching for a reason, astronomers hypothesized that the youngest pulsars had the shortest periods and pulsars slow down as they age. Only young and hence short period pulsars would be found associated with supernova remnants, as the remnants associated with old pulsars whould have dissipated into space, fading from view.

Almost instant confirmation of this idea came from the discovery that the shortest period pulsar, the Crab pulsar, was not completely regular. Within a month the period had been measured accurately enough to demonstrate that it had increased by one millionth of a second. This

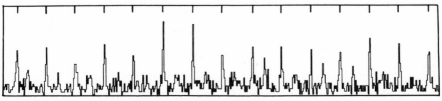

Observations of the light flashes from the Crab pulsar made with the 4-meter Anglo-Australian Telescope. The main flashes (marked on the top scale) are 0.033 seconds apart. Between the main flashes can be seen the so-called interpulse, smaller flashes from a weaker beam on the rotating neutron star, following its main beam by 0.013 seconds. Doubled flashes are relatively common among the faster pulsars.

steady increase meant that the Crab Nebula pulsar was slowing down at a rate which indicated a lifetime of just about 1000 years—near enough the age of the Crab Nebula as determined from the Chinese observations of the supernova of 1054! By the time this was known late in 1968, the four pulsars discovered by Bell and her Cambridge colleagues had been under observation both at Cambridge and Jodrell Bank for more than a year, and their periods were becoming known with greater and greater precision. All four were found to be slowing down, but some 10,000 times more gradually than the Crab pulsar. At the present time 83 of the 149 pulsars now discovered are known to be slowing down—not one is known to be speeding up—further indication that the rotation of pulsars slows as they age.

Seeing the Crab pulsar flash

Although radio observations immediately showed clearly that the pulsar discovered by Staelin and Reifenstein was in the Crab Nebula, its position could not be measured well enough with radio telescopes to determine precisely which star was the pulsar. Indeed, although many attempts had been made to identify the first radio pulsars with particular stars—and there were several false alarms—no radio pulsar had been found to be identical to any optically visible star. These earlier disappointments may explain the casualness of the efforts made by optical astronomers with access to large telescopes to identify the Crab Nebula pulsar. One astronomer used the 98-inch Isaac Newton telescope in Britain, six days after the periodicity of the radio Crab pulsar was announced, to observe light from the central regions of the Crab Nebula, but left his data, which was in the form of punched tape readable only in a large computer, unanalyzed for months with no inkling of the discovery hidden within.

Then a team of three astronomers and physicists at the Steward Observatory in Tucson Arizona used the much smaller 36-inch telescope there in January 1969 to look in the center of the Crab Nebula for light flashes with the radio period. In their experiment W. J. Cocke, M. J. Disney and D. J. Taylor used a small computer to add thousandths of the flashes together, and — this turned out to be a significant difference — they had the results of the computer summation available to them in graphical form then and there, while the experiment was in progress. As the pulses came in from the Crab pulsar, they were displayed on a multi-channel recorder — a glowing screen on which pulses appeared as a hump on a line, growing as the astronomers watched. They were able to sweep the telescope over the central regions of the Crab Nebula to see where the pulses came from and were able to say that they came from the vicinity of the two central stars, the northeast and southwest stars named by Baade. They could not say which, for sure, but guessed the southwest star as this had been the one backed by Baade and Minkowski.

The Steward Observatory experimenters' guess was confirmed in a clever experiment by two astronomers from Lick Observatory, Joe Miller and Joe Wampler. They used a TV camera attached to the 120-inch Lick telescope, peering at the pulsar through a rotating shutter which was made so that the open periods were separated by a time very nearly the same as the period of the pulsar. Similar devices, called stroboscopes, are used to view rapidly rotating machinery; the stroboscopic effect is also familiar as a slowing of the motion of a wagon wheel when filmed by a movie camera which exposes film frame by frame,

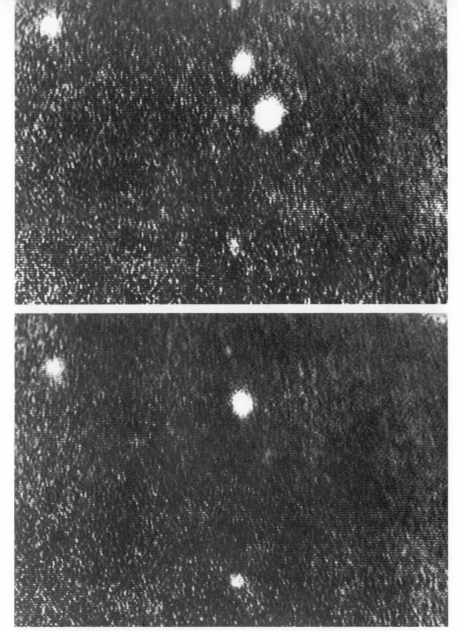

A pulsar flashes. This remarkable pair of photographs of the Crab Nebula pulsar was taken with the Lick 120-inch telescope using a television system looking through a stroboscopic shutter. The top photo was exposed when the pulsar was at its brightest during its flash, the lower when it was almost invisible. The other stars nearby have a constant brightness. This is a Lick Observatory photograph by E. J. Wampler and J. Miller.

through a shutter. When the pulsar was flashing at the same times that the shutter was closed, the Lick astronomers saw nothing. But when it flashed at the times the shutter was open they immediately saw that the pulsar was Baade's southwest star, right at the center of expansion of the Crab, and at the center of activity of the light ripples seen by him.

Astronomers had to look no further for the powerhouse which drives the Crab's expansion and generates the electrons to produce its synchrotron radiation.

A really exciting moment

If a pulsar did not exist in the Crab Nebula it would be necessary to invent one. Indeed it was necessary, for the year before the discovery of pulsars, F. Pacini suggested that a rotating neutron star was the present-day source of the fast electrons in the Crab Nebula. The discovery of the Crab Nebula pulsar showed that this was possible, and the further discovery that the pulsar was slowing down proved it, as the star's spin-energy lost in the slowdown was closely equal to the total amount of light and radio energy radiated by the synchrotron process. The whole flow of energy within the nebula became clearer.

Thomas Gold of Cornell University recalls the moment when he too realized that the Crab Nebula was powered by a rotating neutron star: "Let me just recount for you our excitement at Cornell when we had just obtained, at our observatory at Arecibo, a slowdown rate for the Crab for which we had been looking. We had expected a supernova to give rise to a neutron star, which if found, was expected to be slowing down. Therefore, when the pulsar in the Crab was discovered, we immediately started looking for a slowing down of the pulsations since such a short period pulsar should slow down fast. When we got the information, we immediately worked out the rate of change of energy of a rotating neutron star, having been previously very impressed with the very high total energy content in the rotation of the object.

"I remember doing the completely simple calculation of what the rate of change meant in terms of the energy output, and meanwhile sending an assistant to the library because I no longer remembered the quoted figures for the total energy output of the Crab Nebula that had been calculated by Shklovsky years before. When he returned in a few minutes with various references giving estimates of the luminosity of the Crab we found that the figure on my pad and the other in the book were the same, namely 10^{53} ergs per second. That was a really exciting moment. I realize that we can't be quite sure that this is the right number, but still, it isn't often that you hit it off like that with a completely theoretical calculation of such a far-fetched thing as the structure of a neutron star."

Superstar . . .

There was however still an unexplained problem. The pulsar itself was too faint. Responsible for the vast amount of energy pouring into the Crab and manifest as accelerating filaments and as synchrotron radiation, the pulsar itself emitted just ten millionths as much energy as visible radiation. It was scarcely credible that the pulsar would pulse on its own behalf such a small amount of energy, while passively passing on such huge amounts, any more than it is credible that a showbusiness superstar would accept a small paypacket while generating millions of dollars worth of business. Searching for the Crab pulsar's piece of the action, NRL and Columbia University X-ray astronomers launched two rockets in March 1969 and found X-ray pulsations. This prompted other astronomers to search through and re-analyze old data which they had not previously thought to examine for X-ray pulsations. These pulsations occur exactly in step with the optical flashes, and the period of both is the same.

The energy which the Crab pulsar puts into X-radiation is 20 times the energy it emits optically and 20,000 times the energy it pulses as radio radiation.

... and co-star

Until 1975 the Crab pulsar was the only pulsar detected as an optical or X-ray pulsar as well as a radio pulsar. The search for other examples had proved fruitless. The most likely pulsar to be discovered as an optical or X-ray pulsar was the Vela pulsar in the Vela supernova remnant. Before 1975 it was the second fastest pulsar known and so had more rotational energy to convert to light and X-rays than any other pulsar apart from the Crab. X-rays have in fact been detected from a star near the Vela pulsar, and there was one report in 1973 that the X-rays pulsed, but more sensitive X-ray telescopes failed to see any pulsations. In 1975 however, the Small Astronomy Satellite, SAS-2, detected pulses from the Vela pulsar at very energetic X-ray energies (gamma rays in fact).

Optical astronomers redoubled their efforts to find visible light from the Vela pulsar. Though four times nearer than the Crab and in a less dusty region of the Galaxy, the Vela pulsar was expected to be much fainter than the Crab because ten times older and three times slower. The optical search was hampered by the disagreement among radio astronomers about the precise direction of the pulsar. Within the zone of uncertainty there were too many faint stars to be examined in the detail necessary to pick it up. However after radio astronomers Dick Manchester, Miller Goss and Bruce McAdam had pinpointed the position of the Vela radio pulsar with an accurate radio telescope at Fleurs, near Sydney, Australia, the radio astronomers joined forces with radio-astronomer-turned-optical-astronomer F. Graham Smith, and his team from the Royal Greenwich Observatory, together with four optical astronomers at the Anglo-Australian Observatory, Pat Wallace, Bruce Peterson, John Danziger and Paul Murdin. In 1977, using the newest large optical telescope in the southern hemisphere, the 154-inch Anglo-Australian Telescope, this large team found the Vela pulsar after ten hours integration on the right place in the sky. Light flashes from the Vela radio position were stored in a computer, integrated spontaneously in an analysis at the telescope, and remembered for subsequent more refined analysis afterwards. The Vela pulsar, 10 billion times fainter than bright naked-eye stars, is one of the faintest stars to be seen by optical astronomers.

The Vela pulsar differs from the Crab in that, after accounting for the slowdown of the radio pulse in space, its double gamma ray and optical flashes trail behind the single radio pulse, whereas the Crab emits its radio, X-ray and optical pulses simultaneously. It is too soon to say how this helps us to understand the way pulsars shine.

VII The search for supernova remnants: Looking for other Crabs

When a star goes supernova, it leaves traces. It would be wrong to call these remains the dead body of the star because, in the case of the Crab at least, the remains are still active, emitting as much energy as a luminous star. Are there other supernova remnants? What about Tycho's and Kepler's stars, and all the other historical supernovae?

After Walter Baade had studied the light curve of Kepler's star, he realized that it must have been a supernova and in 1947 he searched for its remnant. Using what was then the world's largest telescope, the 100-inch at Mount Wilson, he photographed the suspect area using a red filter to isolate the red light expected from the hydrogen in the nebula. At the position of Kepler's supernova he found a few wispy filaments of gas, the last traces of the exploding star of 1604.

Baade applied the same principles to a search for the remnant of Tycho's supernova of 1572. He was unsuccessful—photographs centered on Tycho's position for the supernova showed stars but not nebula.

Then in 1952 Robert Hanbury Brown and Cyril Hazard searched for Tycho's supernova remnant with a 218-foot radio telescope at Jodrell Bank (not the fully steerable 250-foot telescope but a simpler telescope fixed to the ground). They found a powerful radio source near the right place, its position being tied down by a Cambridge group using an interferometer. It turned out that the position Tycho gave had the unexpectedly large error of 4 arc minutes—over a tenth of the diameter of the Moon. Though this had caused Baade to look somewhat off the correct place for the nebula, the real position was still on the edge of his plates and he would probably have seen the nebula if it were bright enough.

However, the remnant of Tycho's supernova is very faint indeed and photographs by R. Minkowski with the 200-inch Mount Palomar tele-scope show just a few wisps of nebula near the radio source. The radio source has been mapped by the one-mile Cambridge radio telescope. It turns out to be beautifully circular in shape with a ring of bright radio emission, the edge of a bubble of material thrown violently into the interstellar gas surrounding the supernova, colliding with it, and heating the gas to a billion degrees. This is why so little light is seen—the gas is too hot for hydrogen atoms to form and, in doing so, emit light.

No astronomer doubts that this is the remnant of Tycho's supernova. Why then is it so far from the position Tycho measured for the star? The measurements that he made on the nova with his new sextant extended over many months, but there is no record of Tycho using it for any of his subsequent observations, known to be very accurate. According to astronomer David H. Clark and historian F. Richard Stephenson the discrepancy can be explained purely as a small calibration error in the angular scale on the sextant—as if Tycho were measuring meters with a yard stick.

Once the remains of the supernovae of 1054 (the Crab Nebula), 1572 (Tycho's supernova) and 1604 (Kepler's supernova), had been identified with the wispy supernova remnants, astronomers began to look for other examples. They were hoping to find the remnants of supernovae which had not been recorded, either because they were too faint to be seen or because they occurred in prehistoric times so that no records of the supernovae have come down to us.

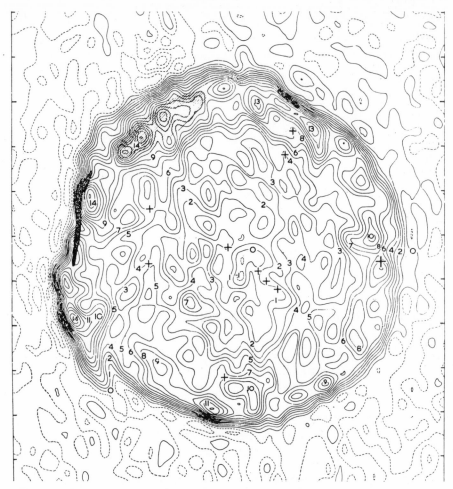

Though the visible remains of Tycho's supernova are faint, this gas shell still glows at radio wavelengths. This is a radio contour map, in which the high number contours refer to peaks of radio brightness, which is greatest around the edge. Hatching shows optically-visible filaments.

The traces of ancient supernovae

In all three cases of the well-established historical super-novae, radio sources were observed at the positions of the supernovae. These radio sources could be distinguished from other radio sources that have been recorded, firstly because supernova remnants have a very characteristic shape and are as obvious to radio astronomers as planetary (ring) nebulae are to their optical colleagues. In fact, the basic causes are the same. When a low-mass star has to shed excess mass, it puffs it off in a shell, a spherical bubble, which is then visible as a ring-shaped planetary nebula glowing in the optical region. A supernova explosion, though much more dramatic, throws off a similar, easily identifiable, radio shell. Second, the radio spectra of supernova remnants are different from those of other radio sources. Radio source spectra are less detailed than the spectra of optical sources. Radio astronomers measure the source's intensity at a number of frequencies, plot a graph of the results, and look at the shape of the curve produced. This varies with the way the radiation is being created. Broadly speaking, there are two types of mechanism for producing radio waves, thermal and non-thermal. Thermal sources are ones in which hot gas is involved: the radiation is produced in the same way as the optical light of stars, by encounters between electrons. In thermal sources relatively slow electrons in a plasma are encountering and mutually repelling one another, changing their motion at random. Whenever an electron changes its motion it emits radiation, so the plasma radiates radio waves with a balanced proportion of long and short wavelengths which is characteristic of thermal sources.

Supernova remnants, however, are examples of non-thermal sources. Relatively fast-moving electrons are colliding not with each other but with the magnetic field embedded in the plasma. In fact the electrons spiral along the magnetic field, emitting radio radiation as they go—the synchrotron process. The spectrum of radio radiation which they emit depends on the fine details of the strength of the magnetic field and the like; therefore although the balance of long and short radio wavelengths is not unique it is usually sufficiently different from the thermal radio sources to be distinguishable.

The latest catalogs of supernova remnants, that is, of sources which are extended and have non-thermal radio spectra, list approximately

100 supernova remnants. The remains of just over two dozen are bright enough to have been detected by optical astronomers as well as radio astronomers.*

The most intense radio source in the sky is in fact a supernova remnant. In the constellation Cassiopeia, it is known as Cassiopeia A, Cas A for short. It was even visible but unrecognized on Reber's first maps of the sky in radio waves, and it was rediscovered in 1948 by Martin Ryle and F. Graham Smith in Cambridge. Smith determined its position accurately and, in 1954, Baade and Minkowski discovered the optical counterpart of Cas A. They found an almost complete shell of knots and filaments approximately four arc minutes across.

Some of these knots and filaments were shown by Baade and Minkowski to be expanding from a central position although no star was visible there, so that there is no neutron star visible. The knots appear to come and go, with only one third of the knots being visible for more than 12 years. They are probably produced by parts of the supernova plowing through stationary interstellar cloud banks. As they strike the cloud banks, atoms in the knots become what physicists term "excited." Electrons in atoms of oxygen, sulfur and argon are moved to high energy states. When they drop back to their original low energy states, they emit spectral lines of these elements.

The ejection velocity and momentum in the knots is so high that collisions do not appear to have slowed the knots significantly since they were first ejected. Some very fast-moving knots are speeding away from the central part of Cassiopeia A at a speed of 9500 km/s. From the expansion velocity, Sidney van den Bergh and William W. Dodd have deduced that the supernova which produced Cassiopeia A took place within eight years of 1667.

The strange thing is that no supernova is recorded in Chinese,

* The twenty-seventh was found by Andrew Longmore, David Clark and Paul Murdin during the preparation of this chapter. In the constellation of the Southern Cross, it has the unexciting name of G296.1-0.8.

TABLE II   *Some supernova remnants*

Name(s)	Supernova date	Distance (l.y.)	Description					
	AD							
Cassiopeia A	1667±	9100	R	O	X			S
Kepler's SN	1604	26,000	R	O		H		S
Tycho's SN, 3C 10	1572	10,000	R	O	X	H		S
3C 58	1151	28,000	R			H		D
Crab, Taurus A, 3C 144, M1, NGC 1952	1054	6500	R	O	X	H	P	D
P 1459-41	1006	4200	R	O	X	H		S
MSH 14-63	185	3000	R	O		H		S
	BC							
Puppis A	2000	7200	R	O	X			S
Vela X	8000	1600	R	O	X		P	S
Cygnus Loop, Veil Nebula	20,000	2600	R	O	X			S
IC 443	60,000	5000	R	O	X			S

Key to description:
R = radio remnant   H = historical supernova
O = optical remnant   S = shell
P = pulsar known   D = disk
X = X-ray emitter

Korean or Japanese chronicles at this date, which suggests that when this Cassiopeia supernova occurred, it was fairly faint. Its distance is comparable with that of Tycho's supernova, namely, just over 9000 light years. It may have appeared fainter than Tycho's supernova because it occurred in a very dusty region of the Galaxy and its light was obscured.

Very detailed radio maps of Cassiopeia A have been made by Ivan Rosenberg, using the one-mile radio telescope at Cambridge. These maps show that Cassiopeia A is a very well defined shell supernova remnant although the shell has broken up into lumps in places. The shell appears donut shaped when seen projected on the sky, because where the line of sight passes along the edges of the shell, we are looking through a greater thickness of gas than at the center.

The force of the collision of pieces of the exploding supernova with the interstellar medium pervading the area nearby heats the gas very strongly, to many millions of degrees. At such temperatures gas emits X-rays and, indeed, these have been detected coming from many supernova remnants (besides the Crab Nebula which, as we saw in Chapter V, emits X-rays by the synchrotron process). The remnants of Cas A and Tycho's and Kepler's supernovae are all in this class. Most

An X-ray "photograph" of the Cygnus Loop supernova remnant, prepared from data obtained in the rocket flight of an X-ray telescope operated by Saul A. Rappaport and his MIT colleagues. The brighter areas (paler shades of gray) represent the most intense regions of X-rays. The remnant has a distinct shell structure and a central source, possibly associated with an unseen neutron star.

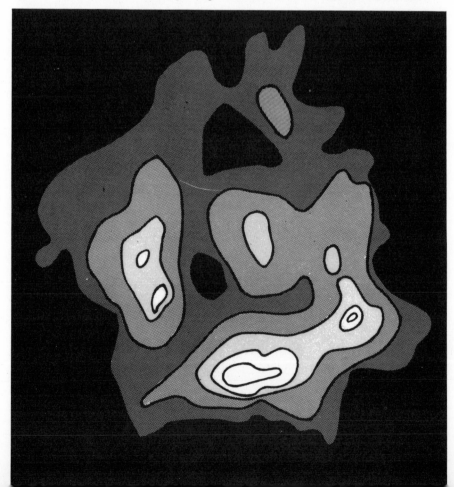

supernova remnants are too small in extent to be perceived as anything but a spot of radiation by the current generation of X-ray telescopes, whose ability to see detail is poor. The largest supernova remnants such as the Cygnus Loop and Vela x, however, can clearly be seen as shells of X-ray emitting gas, roughly similar to the optical and radio pictures. Cas A is probably shell-like at X-ray wavelengths too.

The Gum Nebula and the Vela pulsar

The largest nebula in the sky is called the Gum Nebula. It was discovered by a young Australian astronomer, Colin Gum, whose career was cut short by a fatal skiing accident at the age of 36. For his Ph.D. thesis, Gum had photographed the whole of the Milky Way as it is visible in the southern hemisphere. He used a red filter so that the photographic plate would pick up predominantly the deep red spectral line emitted by hydrogen, H-alpha. Because of the predominance of hydrogen in the interstellar medium, H-alpha is usually the strongest spectral line emitted by nebulae. At the same time, the light of other objects in the heavens, including the glow of our own atmosphere, is fairly weak at the red end of the spectrum. By preventing all but red light from reaching the photograph, it is possible to isolate H-alpha emitting objects from the others, and give longer exposures without white light from other objects flooding the photograph. Gum made a mosaic of several of his photographs of the constellations Vela and Puppis, and the nebula which now bears his name at once became apparent. It has a diameter of about 30 degrees on the sky. This means that if a person's eyes were sensitive enough to perceive the faint H-alpha emissions from the Gum Nebula, it would fill half the area of the sky that one eye can see at a time.

Within the Gum Nebula lie the two stars Zeta Puppis and Gamma Velorum. Gum thought that these two stars were emitting enough ultraviolet light to be able to separate, or *ionize*, any hydrogen atoms in the vicinity into electrons and protons. The result would be that when

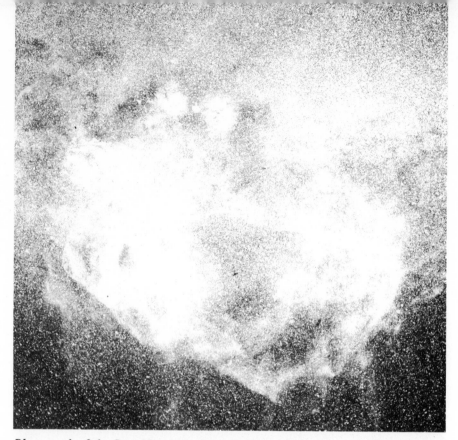

Photograph of the Gum Nebula by John Brandt, Robert Rosen, J. Thompson and D. J. Ludden, made with red light in a wide-angle camera to show H-alpha emission over an area 40 degrees square. The Vela supernova remnant and Vela pulsar are below left of center. Absorption bands of nearby dust (dark zones) obscure the millions of stars of the Milky Way as it runs left to right across the center of the picture, from Carina through Vela, Pyxis and Puppis. The appearance of the Gum Nebula in this picture is consistent with it being caused by the flood of photons released in a supernova explosion.

these electrons and protons recombined into hydrogen atoms, they would emit the H-alpha light that he saw coming from the nebula. The recombined atoms would almost immediately be re-ionized by the two stars' ultraviolet light, to recombine again in a repetitive cycle; the nebula is constantly regenerated. The two stars are 1500 light years away from the Earth, while the Gum Nebula is approximately 1000 light years in radius, so that the Sun lies close to the nearer edge, not quite within the Gum Nebula, but not far away.

In 1971, a group of four astronomers, John Brandt, Theodore Stecher, Steve Maran and David Crawford, calculated that Gamma Velorum and Zeta Puppis did not produce enough energy to be able to split apart the large number of hydrogen atoms within such a large volume of space. They said instead that the hydrogen atoms of the Gum Nebula had originally been split apart, not by the steady radiation from the stars within it, but by a burst of radiation from the explosion of a supernova some 10,000 years ago. The ionized gas will ultimately all

Many thousands of years after a supernova explosion all that remains is a spherical shell. This is the Veil Nebula in Cygnus, one of the brightest supernova remnants. The wisps glow visibly at the edge of a radio shell. Radio astronomers have found no trace of a pulsar at the center of this object, which is about 20,000 years old. Hale Observatories photograph.

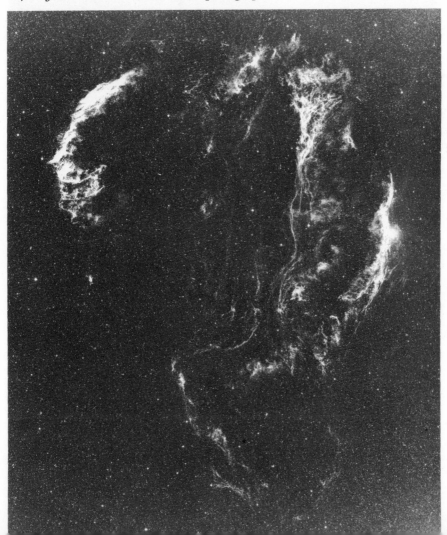

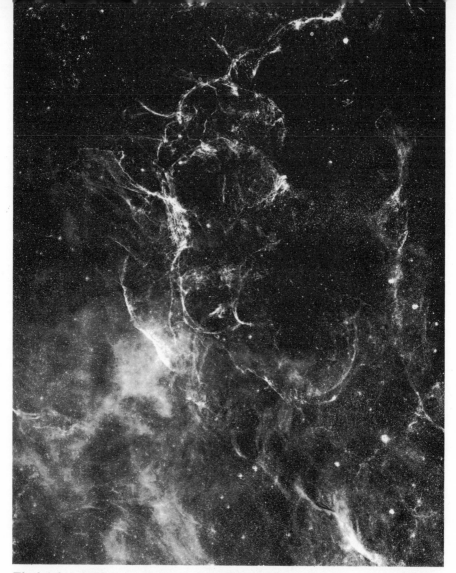

The brighter half of the filamentary shell of the Vela supernova remnant fills most of this picture taken in the red light of H-alpha by John Meaburn and Ken Elliott using the 48-inch UK Schmidt Telescope at Coonabarabran, Australia. The Vela pulsar lies at center right, though far too faint to show. The milky diffuse patches, most prominent at lower left, are parts of the Gum Nebula, recombining hydrogen atoms which were possibly separated by the initial flash of radiation from the Vela supernova some 10,000 years ago.

The motley collection of small fragments in the upper left is the visible remains of another supernova remnant, the strong radio source Puppis A. Science Research Council copyright.

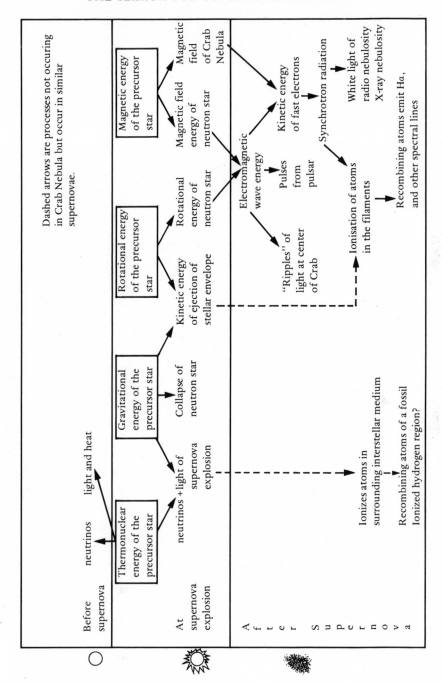

recombine to make hydrogen atoms and the nebula will cease to shine, but the recombination process takes many hundreds of thousands of years because the distance between the electrons and protons is so vast that a wandering electron seldom passes close enough to a proton to be grabbed and made into a hydrogen atom.

According to this interpretation, the Gum Nebula is not a nebula in the usual sense since the stars present are insufficient to keep the nebula constantly ionized. It is a new kind of nebula created suddenly by a supernova explosion and gradually returning to its normal state, not being renewed. The Gum Nebula is a fossil left behind by an event long past.

One must admit that many astronomers are unhappy with the interpretation of the Gum Nebula as the fossil of a relatively recent supernova explosion. Some argue that the Gum Nebula may be a remnant from a supernova that exploded a million years or more ago. They argue that its size has been over-estimated and that its two stars, Gamma Velorum and Zeta Puppis are indeed capable of ionizing the hydrogen within it in order for it to emit H-alpha light.

Is there evidence that a supernova did occur somewhere near the center of the Gum Nebula about 10,000 years ago? The answer is yes. In the center of the Gum Nebula is the smaller nebula known as the Vela x supernova remnant (SNR). This SNR was discovered by Douglas Milne in 1968 using a radio telescope, and we first came across it in Chapter VI, in connection with the discovery of pulsars. Photographs of the area show a filamentary nebula looking just like other well-known SNRs, such as the much-photographed Veil Nebula in the northern sky. A little off-center from the Vela x SNR lies the Vela pulsar. The Vela pulsar's period is just less than $\frac{1}{10}$ second, and is changing at such a rate that it doubles its period every 11,000 years. This can be taken as a measure of the age of the pulsar, which is consistent with the charac-teristics of the Vela x SNR and with estimates of its age from other evidence.

VIII *Types of supernovae: gathering the evidence*

Astronomers are convinced of the connection between supernovae, pulsars, and supernova remnants. But of course a supernova does not appear from nothing. What stars become supernovae? It turns out that there are at least two kinds of stars which produce supernovae: only one kind has been confidently identified. One answer emerged when the supernovae themselves were studied and it was Fritz Zwicky who organized the effort required to determine it. When he first set up the Palomar Supernova Search, Zwicky arranged a cooperative effort to follow the discovery of a supernova by detailed study. Walter Baade would measure the light curves and Rudolph Minkowski and Milton Humason would obtain spectra of the supernovae using the largest telescopes then available, the 60-inch and 100-inch Mount Wilson telescopes. Like other professionally-used astronomical telescopes, these are scheduled in advance, so that a particular night is pre-assigned to a specific astronomer to work on a project whose value he has justified to a committee, in competition with other observers who also wish to have the use of the telescope. Because present-day large telescopes are over-subscribed by a factor of two or three, the competition is fierce. There is not enough time on the telescope to satisfy all astronomers.

By their character supernovae are unpredictable, therefore one cannot pre-assign time on telescopes for their study. Zwicky persuaded the

Mount Wilson Observatory Director, George Hale, to set up an override on observing time on the big telescopes at Mount Wilson such that the scheduled astronomer had to yield time for Baade or Minkowski to observe supernovae bright enough to be worth studying. (An override system for bright supernovae still operates at the 98-inch Isaac Newton Telescope in the U.K. An informal system now operates at Palomar and the scheduled astronomer is not obliged to give way, except insofar as he wishes to continue to have good relations with his colleagues.)

Two kinds of supernovae

From studies of spectra of the first supernovae discovered in Zwicky's supernova search, Rudolph Minkowski found in 1940 that most supernovae discovered (three quarters) are of a single type, which he called Type I. The remainder were different from Type I, though not all alike, and they are known as Type II. (Zwicky has been able to recognize further Types III, IV and V, but these types consist of isolated examples and not all astronomers are convinced that they are distinct kinds of supernovae.)

The Type I supernova has an instantly recognizable light curve, consisting first of the sudden spectacular brightening, a quick decline in less than a month, and then a slower fade-off. In the case of the supernova of 1937 in the galaxy IC 4182, the brightest supernova of this century, astronomers followed this fade-off for more than 600 days during which time the supernova faded without showing signs of stopping its decline. This supernova light curve is considered the prototype of Type I.

Wondering whether the supernovae seen in our Galaxy were of Type I or not, Walter Baade plotted in 1943 the light curves of Tycho's and Kepler's supernovae of 1572 and 1604 on the same graph. He was helped by the almost unbelievably accurate observations of the stars' brightnesses made by Brahe, Kepler and other 17th century astronomers, with no instruments to help them apart from their eyes.

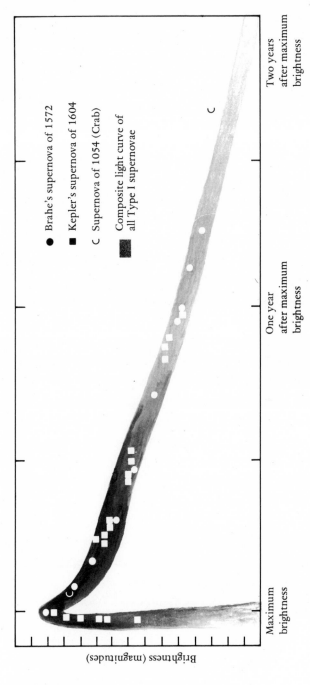

Brightness (magnitudes)

Maximum
brightness

One year
after maximum
brightness

Two years
after maximum
brightness

● Brahe's supernova of 1572

■ Kepler's supernova of 1604

C Supernova of 1054 (Crab)

■ Composite light curve of
  all Type I supernovae

Brightness of both Brahe's supernova of 1572 and Kepler's of 1604 follow composite curve for all Type I supernovae. Only discordant point is for supernova of 1054, raising doubt that Crab supernova was Type I. Few Type I s have been followed for more than a year past maximum.

133

Light curve of Brahe's supernova

Astronomers describe the brightness of a star by a number they call its *magnitude*, but rather confusingly the magnitude scale seems to run the wrong way. Bright stars are said to be of the first magnitude, and the faintest ones which can be seen with the naked eye are sixth magnitude. Venus and Jupiter are brighter than first magnitude stars, and so actually have negative magnitudes. With a telescope one can perceive fainter and fainter stars which have bigger and bigger magnitudes. The faintest star whose light has been detected is the Vela pulsar at magnitude 25. The brightest is the supernova of 1006 which, at magnitude -8 to -10, was 100 million million times as bright as the Vela pulsar.

Brahe's observations of the supernova of 1572 are drawn up in a form similar to that used by present-day astronomers to measure the brightness of variable stars. They cause starlight to fall into a photometer, an instrument which produces an electric current indicating the star's brightness. By pointing the photometer at constant stars whose brightness or magnitude has already been determined and then at the variable star they see how the constant stars match with the variable star. This is just what Tycho Brahe did, except that he used his eye, not an instrument, to compare the brightnesses. A feature of this method is that if at a later date more accurate magnitudes of the constant stars become available, the observations of the variable star can be reconverted into equally accurate magnitudes.

Thus, when Tycho says that in July and August 1573 the nova was equal to the principal stars in Cassiopeia, he himself deduces that it was third magnitude, this being the standard of brightness set down on the basis of those stars' appearance to Ptolemy and published in his book *Almagest* (AD 144). The four stars in Cassiopeia which Ptolemy says are third magnitude are Alpha Cassiopeiae, whose magnitude measured with a photometer is 2.47, Beta Cassiopeiae at 2.42, Gamma Cassiopeiae at 2.25, and Delta Cassiopeiae at 2.80. The average is close to magnitude 2.5, so this was the magnitude of the supernova between July and

August 1573 in the modern system of measuring magnitudes. The fact that the observations have been converted into modern stellar magnitudes 400 years after they were made, rather than the day after, is immaterial and a surprisingly accurate light curve of Brahe's supernova can be drawn up. He wrote:

When first seen the nova outshone all fixed stars, Vega and Sirius included. It was even a little brighter than Jupiter, then rising at sunset, so that it equaled Venus when this planet shines in its maximum brightness. The nova stayed at nearly this same brightness through almost the whole of November. On clear days it was seen by many observers in full daylight, even at noontime, a distinction otherwise reserved to Venus only. At night it often shone through clouds which blotted out all other stars.

However, the nova did not retain this extraordinary brightness throughout its whole apparition but faded gradually until it finally disappeared. The successive steps were as follows:

As already stated, the nova was as bright as Venus in November [1572]. In December it was about equal to Jupiter. In January [1573] it was a little fainter than Jupiter and surpassed considerably the brighter stars of the first class. In February and March it was as bright as the last-named group of stars. In April and May it was equal to the stars of the second magnitude. After a further decrease in June, it reached the third magnitude in July and August, when it was closely equal to the brighter stars of Cassiopeia, which are assigned to this magnitude. Continuing its decrease in September, it became equal to the stars of the fourth magnitude in October and November. During the month of November, in particular, it was so similar in brightness to the nearby eleventh star of Cassiopeia that it was difficult to decide which of the two was the brighter. At the end of 1573 and in January 1574 the nova hardly exceeded the stars of the fifth magnitude. In February it reached the stars of the sixth and faintest class. Finally, in March, it became so faint that it could not be seen any more.

Walter Baade condensed this description to modern magnitudes and produced Table III. He then turned to Kepler's supernova of 1604.

TABLE III *Magnitudes of Tycho's supernova*

Date	Days after maximum brightness	Description	Magnitude
1572			
Nov	0	Almost as bright as Venus	−4.0
Dec	30	About as bright as Jupiter	−2.4
1573			
Jan	61	A little fainter than Jupiter	−1.4
Feb–Mar	107	Equal to brighter stars of first mag	+0.3
Apr–May	167	Equal to second magnitude stars	+1.6
Jul–Aug	259	Equal to Alpha, Beta, Gamma, Delta Cassiopeiae	+2.5
Oct–Nov	351	Equal to stars of fourth magnitude	+4.0
Nov	365	Equal to Kappa Cassiopeiae	+4.2
1573			
Dec–Jan	412	Hardly brighter than fifth mag	+4.7
1574			
Feb	457	Equal to stars of sixth magnitude	+5.3
Mar	483	Became invisible	+6.0

Light curve of Kepler's supernova

The light curve of Kepler's supernova can also be well determined from the 17th century observations because Mars, Jupiter and, later, Saturn, provided useful comparisons. The maximum brightness was somewhat brighter than Jupiter in mid October 1604, and it was still almost as bright as Jupiter when it disappeared near the Sun in November. In January, when it reappeared, it was only about as bright as Antares. It continued to become fainter until it reached the fifth magnitude in October 1605. By the following spring, it was no longer visible to the naked eye. Throughout its appearance Kepler made a series of comparisons between it and other stars, possibly modeled on Brahe's observations of the supernova of 1572.

TABLE IV *Magnitudes of Kepler's supernova*

Date	Days after maximum brightness	Description	Observers	Magnitude
1604				
Oct 8	−9	Not seen	(several)	+3 or more
Oct 9	−8	As bright as Mars	The physician	0.9
Oct 10	−7	Like Mars	Capra Marius	0.5
Oct 11	−6	Still brighter than Oct. 10	The physician	−0.7
Oct 12	−5	Almost as bright as Jupiter	Roeslin	−1.5
Oct 15	−2	As bright as Jupiter	The physician and Fabricius	−2.2
Oct 17	0	Much brighter than Jupiter	Kepler	−2.6
1605				
Jan 3	78	Brighter than Alpha Scorpii	Kepler	0.9
Jan 13	88	Brighter than Alpha Bootis and Saturn	Kepler	0.0
Jan 14	89	About as bright as Mars in Oct 1604	Fabricius	0.9
Jan 21	96	As bright as Alpha Scorpii	Mästlin	1.2
Jan–Feb	100	As bright as Alpha Virginis	Heydon	1.2
Mar 20	156	Not much brighter than Zeta and Eta Ophiuchi	Kepler	2.4

137

Date	Days after maximum brightness	Description	Observers	Magni-tude
Mar 27	163	Same	Brengger	2.4
Mar 28	164	Not much brighter than Eta Ophiuchi	Cristini	2.4
Apr 12	179	As bright as Eta Ophiuchi	Fabricius	2.6
Apr 21	188	Same	Kepler	2.6
Aug 13	302	As bright as Xi Ophiuchi	Kepler	4.5
Aug 29	318	About as bright as Xi Ophiuchi	Kepler	4.5
Sep 13	333	Fainter than Xi Ophiuchi	Kepler	5.0
Oct 8	356	Fainter than Xi Ophiuchi, difficult to see	Kepler	5.8

Drawing the light curves of Tycho's and Kepler's supernovae on the same graph as the light curve of the supernova he observed in IC 4182, Baade found that all three objects are the same kind of supernova—Type I.

The Crab—Type I or II?

About the supernova of 1054, the Crab Nebula supernova, there is, however, some doubt. The data are scanty but it seems that after SN 1054's fairly quick decline to magnitude -3.5 it faded to magnitude $+5$ in 630 days, a drop of 8.5 magnitudes, compared with the norm of 10.5 magnitudes for the prototype Type I. Thus, writes Minkowski, "the difference of 2 mag between the supernova of 1054 and supernovae of Type I is not conclusive evidence in view of the many uncertainties, but it tends to contradict the interpretation of the supernova of 1054 as Type I and certainly does not make it mandatory". This is important in trying to explain why the remnants of Tycho's and

Kepler's supernovae are so different in their appearance from the Crab Nebula: it would be easier to understand if the former were Type I and the latter Type II.

The true brightness of a supernova

The brightness of all Type I supernovae at their maximum is probably the same, insofar as this can be tested. There are three uncertainties which have to be considered. The first is that many supernovae are on the decline when first discovered and there is no way of determining what their maximum brightness is. Only a supernova whose maximum brightness is well established can be used in this test.

Second, supernovae are at varying distances from Earth so that their brightness as measured depends on how far away they are. Astronomers have to correct their measurement of the brightness of the supernova by calculating how much brighter it would appear if placed at some standard distance from Earth. (For various historical reasons this standard distance has been chosen as ten parsecs, about 32.5 light years.) To make the calculation they have to know at what distance the supernova lies. This can be done for the nearest galaxies by comparing the brightness of certain kinds of variable stars in those galaxies (cepheids) with cepheids in our Galaxy. Often, however, no individual stars can be distinguished in a galaxy in which a supernova occurs (save the supernova itself) and astronomers fall back on the so-called Hubble redshift relation.

Edwin Hubble found in 1929 that on average all galaxies were receding from our Galaxy at speeds such that the more distant galaxies recede faster, and their spectra are redshifted more, because of the Doppler effect. According to the best recent determinations 55 km/s is added on average to a galaxy's recessional speed for every million parsecs (3.25 million light years) distance it is from Earth. Thus, if a galaxy's speed is measured by the Doppler effect to be 5500 km/s, its distance is around 100 million parsecs. A supernova of Type I might have a maximum brightness of 16 in this galaxy, but when brought to a distance of just 10 parsecs (10 million times closer) it would be 10 million squared times brighter, that is 100 million million times brighter. Each factor of 100 in brightness corresponds to 5 magnitudes, so 100 million million means a difference of 35 magnitudes, and the supernova would appear magnitude $16 - 35 = -19$ if it were at a distance of 10 parsecs—far brighter than the Full Moon.

The third correction that ought to be made is for the amount of light from the supernova that is absorbed by dust in our Galaxy, and in the parent galaxy of the supernova. The absorption correction corresponding to our own Galaxy's dust can usually be made with adequate accuracy on the simple assumption that the dust forms a slab in our Galaxy, parallel to the Milky Way, through which astronomers point their telescopes towards supernovae in different directions, looking through different slanting thicknesses of dust. Astronomers have no information on the absorption in the parent galaxy, because they cannot tell whether the supernova is on the nearer or farther side of the galaxy.

The latest result of these calculations on the absolute magnitude of Type I supernovae is that the average magnitude is -19.5 to -20, corresponding to a luminosity nearly ten billion times that of the Sun. This number is considerably in excess of the first estimates (in the 1930s) of the luminosity of supernovae because at the time the scale of the Universe was considerably underestimated.

Knowing now that the absolute magnitude of Type I supernovae is about -19.5, we can attempt to calculate how far away such a supernova has to be pushed in order to appear magnitude -4.0 (Tycho's supernova) or -2.6 (Kepler's supernova). These calculations are somewhat handicapped by the correction for dust in our Galaxy. Since these two supernovae are within the galaxy, the assumption that the absorption lies in a slab is no longer relevant (the supernovae might be within the slab). We can, instead, use the fact that dust in the Galaxy not only absorbs starlight, it reddens it too, just as the Sun is reddened at sunset when its sunlight passes slantwise through the Earth's atmosphere. If we can obtain information about the color of the two early supernovae

and know what color (by present-day measurement of Type I supernovae in other galaxies) they might have appeared had there been no
dust, then we can determine how much dust lies between Earth and the
supernovae. We have to rely on subjective impressions by Tycho and
Kepler of the color of the respective supernovae and the argument
treads on quicksand at this point.

However, all observers stressed that at its maximum Tycho's supernova had the yellow color of Venus or Jupiter, becoming redder like
Mars or Aldebaran after a month and whiter like Saturn at the same
time thereafter. Kepler's supernova seems to have been even redder on
the whole than Tycho's. On the basis of a discussion of these observations, Minkowski concluded that the absorption of light from Tycho's
and Kepler's supernovae by dust in our Galaxy was 2.1 and 3.3 magnitudes respectively. Thus, had there been no dust, they would both have
appeared at magnitude -6, or 14 magnitudes fainter than if at the
standard distance of 32 light years. They are both, therefore, some
20,000 light years distant. We have to be wary of the large uncertainty
in this estimate, caused mainly by the doubt about the reddening
correction.

What about the light curves of the Type II supernovae? There is
considerable diversity in these but after the maximum, Type IIs typically fade by a magnitude and a half and then almost halt their decline
for 50 days. They then decline more rapidly at no particular rate, fading
by two magnitudes in 60 to 120 days. Because Type IIs are more
diverse, because they are rarer than Type Is, and because few have been
well observed, their absolute magnitude is not known with certainty,
but there is a suspicion that they may be fainter than Type Is by a
magnitude or two.

Spot the difference

The original distinction between Type I and Type II
supernovae was made by Minkowski by looking at their spectra. By 1941
he had taken spectra of 14 supernovae and found that nine were very

similar in showing various distinctive bands of colors, whereas spectra of the five examples of the second type were featureless during the period of maximum brightness and then developed distinctive features because of hydrogen. The Type IIs look most like common novae— their spectra have blue-shifted spectral lines, indicating that material is approaching Earth at speeds as high as 15,000 km/s at the time of the supernova outburst. Presumably these lines are caused by a shell of material ejected from the supernova, which of course on its near side will be approaching Earth. Type Is have less understandable spectra, but recently it has been established that they have the same basic cause, though the expansion speed of the shell of material reaches some 20,000 km/s.

The difference between Type I and Type II supernovae is not just the distinction between two engravings of the same postage stamp, a fascinating exercise with little meaning beyond itself. The clue to the significance of the difference is given by the kind of galaxies in which the two types of supernovae are found.

Not all galaxies are alike. Once Hubble had established that galaxies were beyond the Milky Way (extragalactic) he went further. From the vast numbers of photographs which he accumulated, he established the existence of a series of galactic shapes, ranging from irregular con- glomerations of dust, gas and bright young stars, through spiral galaxies with arms in which the dust, gas and bright young stars concentrate, to structureless spherical or elliptical balls of old, red stars with scarcely a trace of dust and gas. Although Hubble himself warned against the implication that this sequence of galaxian forms represented the evolutionary life of a galaxy, it is now thought that there is some connec- tion between the appearance of a galaxy and the age of the stars within it. Only in spiral galaxies and in irregular galaxies is there still a supply of dust and gas to form young stars now. For some reason star formation in elliptical and spherical galaxies has long ceased. The massive bright young stars in elliptical galaxies have all ceased to shine and only stars like the Sun still visibly survive in them.

When astronomers examine the frequency with which supernovae occur in various kinds of galaxies, they find that no Type II supernova has ever occurred'in an elliptical galaxy, or even in the kinds of galaxies considered as intermediate in shape between the ellipticals and the spirals. Only in galaxies with clear spiral arms have Type IIs ever been seen. Type Is on the other hand appear in all kinds of galaxies but favor the elliptical galaxies. The implication is that Type II supernovae are the consequence of the evolution of the more massive stars (with masses, say, of ten times the solar mass) whereas Type Is are caused by the evolution of stars typically of one solar mass (or less).

While most astronomers are happy with this notion about the precursors of Type IIs, many have expressed qualms about accepting the contrary implication for Type Is. Some have proposed that these come from exploding white dwarf stars or from close double stars. Some have claimed that there must be a few very massive stars in elliptical galaxies, though not as many as in spirals, and that Type Is, like Type IIs, are both formed from massive stars. The origin of the more frequent type of supernova, the Type Is, remains obscure.

IX The making of a neutron star: what makes pulsars tick

The stars, like measles, fade at last.
Samuel Hoffenstein

So far we've seen how astronomers have demonstrated observationally that massive stars produce supernovae and that supernovae make neutron stars. Knowing that this occurs is not enough; astronomers want to know why. To understand why we have to look at the life history of stars in some detail.

The life and death of stars

The modern theory of stellar evolution goes like this. All over the Galaxy there are clouds of interstellar hydrogen, vague, tenuous, gaseous masses, called *nebulae* when they can be seen shining, which they do if they are near a bright star. For reasons which are not entirely clear, especially dense parts form within a nebula, and these dense parts have a sufficiently strong mutual gravitational attraction to make them fall together in larger and larger lumps called *protostars*. They are "about-to-be" stars. Gravity is the force which attracts these lumps to one another, just as the force of gravity on Earth pulls all objects down to the surface, drawing them inexorably towards the center.

As the gas of which the protostar is made is packed closer and closer in towards its center, it begins to heat up, just as the air in a bicycle pump heats as it is compressed into the tire. You can feel the valve getting hot

144

as you pump. Such heating causes the star to stop collapsing. It causes the pressure in the center of the star to increase, and this pressure shoulders the atoms of gas apart which in turn keep the outer layers from collapsing. The star becomes a finely balanced mechanism, tending to collapse into itself because of its own gravitational force, but prevented from doing so by the pressure of the gas inside.

This balance comes about only when the gas in the center of the star reaches a temperature of many millions of degrees. In a star like the Sun, the center reaches a temperature of 15 million degrees centigrade and a density of about 160 times that of water. In the innermost three per cent of the volume of a star like the Sun, into which is packed two thirds of the mass of the star, the temperature and the density are so high that nuclear reactions take place. Over the whole star the hydrogen atoms from which the star is made have been split apart into their component bits by the force of the collisions between the atoms as a result of the high temperature; the hydrogen atom is so simple that it produces only two pieces, an electron and a proton, its massive central nucleus. In the very center of the star the protons themselves have been forced so close together that there is a substantial chance that four protons will be able to stick together in a new arrangement called an alpha particle which is the central massive part, or nucleus of a helium atom. As the four hydrogen nuclei are converted into a helium nucleus, energy is released, mostly in the form of gamma rays, which are very energetic light-particles or photons, similar to light but with a million times more energy.

The gamma rays slowly diffuse up out of the star, being degraded on the way to a much larger number of lower energy photons which ultimately leave the surface of the star in the form mostly of light and infrared radiation. It is this light which leaves the star, travels through space, and permits stars to be seen. Indeed, sunlight is the result of gamma rays emitted from the center of the Sun millions of years ago, released by the nuclear reactions, and providing a fossil record of these.

The energy production within a star can be thought of as brilliantly

145

shining proof of Einstein's famous equation $E = mc^2$ —that is, mass and energy are equivalent, and one can be converted into the other given the right conditions. Each time four protons are converted into a helium nucleus, 0.7 percent of the mass of the four individual protons is converted into radiative energy. Although each nuclear event converts just a small amount of mass to energy, in total a star like the Sun radiates four million tons of its mass away every second of its life.

At this rate of consumption even a mass as large as that of the Sun must be appreciably diminished in the course of time and less hydrogen is available to sustain the rate of energy production. The star becomes unbalanced, unable to support itself against its own gravitational attraction. The center of the star begins to contract and the outer parts begin to expand, producing what is technically known as a *red giant*.

In a sunlike star, this occurs at an age of ten billion years. The conversion of the four hydrogen protons to a helium nucleus still takes place, not in the center but in a shell around the center of the red giant. The center of the star, which is by now largely helium, releases energy first by contracting, but as it contracts, if it is large enough, it eventually becomes hot and dense enough for a further kind of nuclear reaction to begin. The star begins to convert the helium nuclei into carbon nuclei by what is known as the *triple-alpha process*, because it involves the uniting of three helium nuclei or alpha particles to make a carbon nucleus.

This further conversion of energy in the triple-alpha process halts the contraction of the center of the star for a while, but not for long. Because the triple-alpha process is a much less efficient means of generating energy than the conversion of hydrogen to helium, the star remains a red giant for only a comparatively short time. The star's central regions contract even more, and if the star is massive enough, the contraction causes the center to heat up enough for yet further nuclear reactions to occur. In these, helium nuclei are successively added to carbon nuclei to form heavier nuclei like oxygen, neon, magnesium and silicon, possibly up to iron nuclei. But less and less energy is available from these reactions and they only briefly postpone the ultimate collapse of the center of the star to a very dense state, known as *degenerate matter*, when the red giant star turns into a *white dwarf*.

Because it has collapsed so much, the density of a white dwarf is very high: about one million grams per cubic centimeter, so that a block one centimeter on the side—roughly the size of a sugar lump—weighs one ton. Left to itself, the white dwarf gradually cools off, rapidly at first from white hot to yellow hot, but then more slowly to red hot. As it becomes redder, it fades until, ultimately, it disappears from view. At the last stage, it is known as a *black dwarf*.

In 1932 the Indian-born American astrophysicist S. Chandrasekhar proved a remarkable theorem about white dwarfs, showing that no white dwarf could exceed a mass of about one and a half times the mass of the Sun. A star of larger mass which attempted to become a white dwarf would not be able to hold itself up against the force of gravity and would have to turn into something other than a white dwarf. This maximum mass for a white dwarf is known as the *Chandrasekhar Limit*. The Chandrasekhar Limit lies between 1.44 solar masses and 1.76 solar masses depending on precisely which nuclei the star is made of. If it is composed of helium nuclei, the maximum possible mass is 1.44 solar masses, and if of iron nuclei, 1.76 solar masses. Yet the number of white dwarfs known is too large for them to have evolved only from stars which are less than the Chandrasekhar limiting mass. (Near the Sun there are about five white dwarfs in every cube of space whose side is 30 light years.) Somehow, stars which are heavier than the Chandrasekhar limiting mass must lose material before they can become white dwarfs.

How weighty stars lose mass

Stars can lose material at two stages in their evolution. The first is while they are red giants. Because a red giant is very extended, its mass is spread over a large volume and it has a low surface gravity. Bits of the atmosphere can, relatively easily, be thrown back into space by various storms, flares and winds from the star's surface. Then more

147

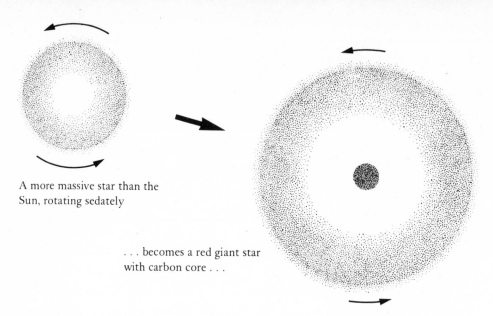

A more massive star than the Sun, rotating sedately

. . . becomes a red giant star with carbon core . . .

dramatic ejection of material into space occurs as a red giant attempts to become a white dwarf, during which it puffs off a layer of material from its surface. As this sphere of material slowly expands away from the star, it becomes so huge that it can be photographed by astronomers, who call it a *planetary nebula*. This is because when seen through a small telescope, it has the appearance of a flat disk like the planets Uranus and Neptune.

A typical planetary nebula is comparable in size to the solar system, but the largest can be a light year or more in diameter. Its mass of around two tenths that of the Sun expands into space at a speed of about 20 kilometers per second. In a time of several thousand years, the nebula disperses into space. Perhaps, in order to become a white dwarf, a massive red giant has to go through the stage of ejecting a planetary nebula several times in order to bring its mass below the Chandrasekhar Limit. Through this process, stars as large as four times the solar mass can become white dwarfs by ejecting planetary nebulae.

Stars larger than about four solar masses cannot cope in this way with the problem of turning quietly into white dwarfs—the forces on them are too extreme. In this case, what exactly does occur is not clear, but it is some kind of catastrophe. Because of the successive "burning" of heavier and heavier nuclear fuels, the central part of the star is made of a series of concentric shells of different nuclei, like an onion, the inner shells containing the heavier nuclei. One of the layers, of carbon or oxygen, explodes and drives the layers outside it off the star. The core inside the exploding layer implodes ("exploding inwards," like a fractured light bulb), and shrinks so that its density increases beyond even

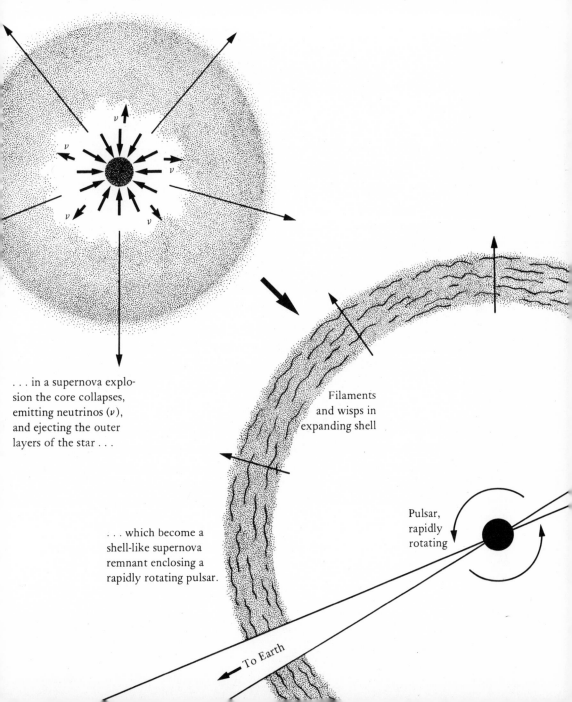

. . . in a supernova explosion the core collapses, emitting neutrinos (ν), and ejecting the outer layers of the star . . .

Filaments and wisps in expanding shell

. . . which become a shell-like supernova remnant enclosing a rapidly rotating pulsar.

Pulsar, rapidly rotating

To Earth

the degenerate density of a white dwarf. It becomes, in fact, a *neutron star*, since electrons in the core of the star are forced into the protons present in the carbon nuclei. The negative electric charge of the electrons cancels out the positive charge of the protons, giving electrically neutral particles called *neutrons*.

The material outside the degenerate carbon core must all be blasted off the star in the ferocious explosion. An explosion of this kind is what astronomers see as a supernova. The explosion ejects most of the mass of the precursor star. Neutron stars are subject to a limit on their mass similar to the Chandrasekhar limit on the mass of a white dwarf. There is some discussion about the size of the critical mass of a neutron star, but C. E. Rhoades and R. Ruffini of Princeton University have calculated that the critical mass certainly cannot exceed 3.2 times the mass of the Sun. All neutron stars have masses below this figure, probably below 2.5 solar masses. A typical neutron star has the same mass as the Sun. In the explosion of a five solar mass star, therefore, three or four solar masses must be ejected if a neutron star is to be the result.

The density of neutron stars is extremely great, at least ten million million grams per cubic centimeter, so that a cube the size of a sugar lump would weigh ten million tons, comparable with all the materials used to build a city the size of London or New York. This is because material to the amount of about one solar mass is packed into a star whose diameter may be only 20 kilometers.

How spinning stars speed up

A curious thing happens to the rotational period of a star which collapses into a smaller volume. All stars rotate to a greater or lesser extent. The Sun, for example, spins once on its axis every 25 days 9 hours, just as the Earth spins on its axis in one day. Stars more massive than the Sun generally rotate faster, and periods as short as half a day are known for some stars. When stars expand and become red giants, their period of rotation lengthens: they slow down. During their subsequent collapse to white dwarfs or neutron stars, they speed up again.

The reason for this is a physical law called the *conservation of angular momentum*, which says that in this kind of situation the cross sectional area of a star divided by its rotational period will remain roughly a constant. When the star expands in the red giant stage its cross sectional area gets larger and therefore its period will become proportionately greater—the star slows down. When the star subsequently shrinks, its cross sectional area becomes smaller, its period is reduced in proportion—and the star speeds up.

Ice skaters exploit the law of the conservation of angular momentum when they pirouette on the ice with arms outstretched, giving themselves a large cross sectional area, and then rapidly bring their arms to their sides. This decreases their cross sectional area and thus decreases their period of rotation; in other words, they spin faster. Consider a star like the Sun with a cross sectional area of 600 billion square miles and a period of about one month (two million seconds). If it were possible to collapse the Sun suddenly to the size of a neutron star, the cross sectional area would decrease by a factor of a billion so that the period too would have to decrease by a factor of a billion, to about a five hundredth of a second. In a real supernova explosion not all the star collapses to a neutron star and the parts which are ejected carry off some of the angular speed of rotation so that in practice neutron stars rotate more slowly than this—their periods are several times longer although still almost incredibly fast. The Crab Nebula pulsar, when it started to rotate, must have been spinning about twice as fast as it is now, with a period of about a sixtieth of a second.

The rotation of heavy objects makes a very good repetitive clock. After all, the Earth was used as a clock until very recently when atomic clocks revealed the small irregularities in its rotation. The regular rhythm of a pulsar, and the Crab Nebula pulsar's rate of repetition of a few tenths of a second, led Thomas Gold to suggest that pulsars were rotating neutron stars. In some way, he suggested, a pulsar emitted a beam of light like a lighthouse, and this beam, as it turned towards the Earth with each rotation, produced the pulses observed by radio and optical astronomers. When he found that the Crab pulsar was slowing down, feeding its pulses from its rotational energy, he became convinced that his idea was correct.

How pulsars may shine

It is not quite enough to say that pulsars are like lighthouses, beaming their light and radio rays into space as they spin around. How actually do they make the beam? The answer is unclear but tied up with an immense magnetic field at the surface of a pulsar. All stars have magnetic fields to a greater or lesser degree and when a star produces a pulsar its magnetic field is caught fast and throttled, squeezed into the smaller cross section of the neutron star. When this happens the strength of the magnetic field increases enormously, by a factor of millions.

Pulsar astronomers assume that the magnetic field of the pulsar does not lie exactly along the polar axis of the pulsar. As in the Earth, whose magnetic axis leaves its surface in Canada, 15° of latitude from the North Pole, the magnetic field of a pulsar may lie off the pole of rotation, possibly even at the pulsar's equator. The pulsar is therefore a powerful electric generator, spinning a magnetic field lying across a rotor, and an electric current pours from the surface of the neutron star. Caught in the magnetic field the electrons which make up the current stream out of the magnetic poles radiating as they do so. The beam of radio waves which they emit points out of the magnetic axis of the pulsar and as the magnetic axis sweeps in the direction of the Earth we perceive a pulse. Possibly we are in a position to see a diminished amount of radiation from the magnetic pole on the far side of the pulsar and then we can see two pulses: many pulsars, including the Crab, show two pulses during every rotation.

Perhaps not all pulsars shine in this way. Perhaps in some the beam of radiation is produced by electrons bunched, not at the magnetic poles, but off the surface of the neutron star near to the place where the magnetic field is moving at speeds close to the speed of light. Then the radiation is caused not by electrons beaming along the magnetic field but along the direction in which they are moving, at right angles to the magnetic field.

Whatever the cause, investigation of the way pulsars shine has stimulated research into the fundamental physics of fast-moving strong magnetic fields and particles. Only in the cosmic laboratory has it been possible to see such an experiment in operation.

Starquakes

Although pulsars are very good clocks, rotating very regularly apart from a gradual slowdown, most pulsars, particularly the two fastest, the Crab Nebula pulsar and the pulsar in the Vela supernova remnant, do, when examined closely enough, show irregularities. At the end of September 1969, for example, the period of the Crab pulsar suddenly decreased by a third of a billionth of a second. There have been three much larger jumps in the Vela pulsar's period, each of 200 billionths of a second. The reason for these sudden so-called "glitches" seems to be that neutron stars have a crust, just as the Earth does. Because they are rotating so fast the crust is not precisely spherical but is flat at its poles and bulges at the equator, due to the centrifugal force there, just as the Earth is tangerine shaped. As the pulsar slows down the centrifugal force becomes less, but the crust takes the strain for a time even though the equatorial bulge of the pulsar tries to fall. The crust cannot survive the stress indefinitely. It breaks and suddenly drops. By the law of the conservation of angular momentum previously mentioned, the pulsar suddenly spins faster.

The amount by which the crust drops is minute—it is measured in millimeters! But this is enough to have a noticeable effect on the pulsar's period just because it is usually so regular.

Astronomers call the sudden changes in the neutron star "starquakes." The energy released in a starquake is enormous and flows out from the star into the space surrounding it. After the 1969 starquake on the Crab pulsar a wave of energy was seen to flow outwards, rippling through the center of the Crab Nebula. This, apparently, is the cause of the waves of activity which Baade noted and was one of the clues which convinced him correctly in 1945 that he had identified the Crab Nebula supernova's stellar remains.

X   *Neutrino astronomy: the ultimate cause of supernovae*

O dark dark dark.
They all go into the dark,
The vacant interstellar spaces,
The vacant into the vacant.
                    T. S. Eliot

Most things in science which we know about have
been *discovered* by somebody. Yet it is possible to talk about some things
actually being *invented*, in the sense that a theoretical scientist saw that
logically they had to exist before they had been discovered. This is so in
the case of the neutrino, which was invented by Wolfgang Pauli in 1930.
The story of the neutrino is also a good example of the way in which the
behavior of sub-microscopic particles, individually all but undetectable,
can have devastating effects on a large scale. For, apparently, neutrinos
can actually cause a supernova.

Pauli had to invent the neutrino because of a fundamental law of
science: energy can be changed from one form to another, but it never
appears or disappears. Atomic scientists in the 1920s were worried that
this law was apparently being broken in a certain kind of nuclear trans-
formation called beta-decay, which is, essentially, the way in which a
neutron spontaneously decomposes into a proton and an electron. A
proton is positively charged and an electron is negatively charged, while
a neutron has no charge at all. Consequently, the result of a neutron
suffering beta-decay is still no net electric charge, since the charges on
the electron and the proton cancel each other. Therefore, electric charge
did not appear or disappear in beta-decay: in the jargon of nuclear

155

scientists, it was conserved. But although the charge was conserved, the problem was that the total energy of the neutron alone was greater than that of the pieces after beta-decay. Energy was not being conserved: it seemed to disappear.

To get over this difficulty, Pauli invented a particle which could have no charge, but which would balance out the energy equation. A small fraction of the energy in beta-decay, he said, was carried off by this imagined particle which no one had ever detected. Because the energy carried away was sometimes very small, the neutrino had to have a very small mass—even zero! Pauli was aware that he was dangerously near sophistry. He thought that no one would ever detect the neutrino and told the astronomer Walter Baade, "Today I have done the worst thing for a theoretical physicist. I have invented something which can never be detected experimentally." Pauli originally called his imaginary particles "neutrons." But they were different from what we now call neutrons, which were not actually discovered until 1932. Enrico Fermi, with exasperation and gestures to match, explained the difference to an audience of slow physicists at a conference in 1933: "The neutrons discovered by Chadwick are big. Pauli's neutrons were small. They should be called neutrinos." The -ino ending in Italian is a diminutive, like *bambino*, and the name stuck.

Fermi worked out that the chances of a given neutrino reacting with anything were very small. If a neutrino travels at the speed of light through a 3000-light-year-thick slab of matter with the density of water (the average density of the Sun), it has only a 50-50 chance of reacting with a proton.

Nonetheless, neutrinos are given off in large numbers from nuclear reactors: ten trillion every second pass through a square centimeter near the reactor. Although each one has only a small chance of being detected in experiments, there are so many available that just a few actually are. Pauli was too pessimistic—the existence of neutrinos has been confirmed.

Now, stars like the Sun are vast nuclear reactors: the Sun creates in its center vast numbers of neutrinos every second. Because the Sun's radius is only 2 light seconds, very small compared with 3000 light years, nearly all the neutrinos created by the Sun dash unheedingly out of the Sun's surface and diffuse through space. By the time they reach Earth, the number passing through each square centimeter per second has been calculated to be 65 billion: a much smaller number than can be made in a terrestrial nuclear reactor, but still just enough to give a few detectable interactions.

Detecting the undetectable

How can scientists detect a particle which will hardly interact at all with matter, passing right through the Earth without being affected? In 1946 B. Pontecorvo, an Italian immigrant to the U.S. who was later to defect to the U.S.S.R., suggested a method by which neutrinos might be found. Pontecorvo's idea is now being realized in an experiment performed under the leadership of Raymond Davis of Brookhaven National Laboratory.

The detector consists simply of a tank containing 610 tons of tetrachloroethylene, a fluid used in the drycleaning trade. Each molecule of tetrachloroethylene consists of four chlorine atoms coupled to two carbon atoms. Approximately once per week, one neutrino out of the expected 120 thousand million million million reaching the tank from the Sun interacts with a neutron in one of the chlorine atoms in the tank. The result is that the chlorine atom, which previously contained 17 protons and 20 neutrons, becomes an argon atom containing 18 protons and 19 neutrons.

After a while, the tank is bubbled for a day with helium which sweeps up the scores of argon atoms created by the solar neutrinos. They are recovered by freezing the argon out over charcoal. The created argon atoms are not stable: they suffer beta-decay on a time scale of a month and eject electrons which can be detected electronically. The strength of the signal detected in this way depends on the number of argon atoms which have been created in the tank since the last run.

157

In practice there are spurious events occurring in the tank, caused by the radioactivity in nearby rocks and by cosmic rays. The cosmic ray contamination has been minimized by constructing the tank one mile below the surface in a gold mine in S. Dakota, so that most of the cosmic rays are absorbed by the mile of rock. Nevertheless, the presence of the spurious events has masked events caused by solar neutrinos.

Before Davis began his experiment, calculations based on the assumed internal temperature of the Sun gave a value for the number of neutrinos to be expected. Expressed in terms of Solar Neutrino Units, the value was 6 SNU.

What SNU?

The actual results from Davis's experiment came as the biggest shock to astrophysicists for many years. As his experimental procedures became more refined, it became clear that the neutrino rate was actually much, much lower than predicted, with at most a level of 1 SNU, around the level of the spurious events. In fact, it can be said that no solar neutrinos have been detected with any confidence at all. This moved Willy Fowler, one of the leading astronomers in this field, to remark at a conference "What SNU?"

During 1976, somewhat higher rates were detected. Davis himself put this down not to a change in the Sun itself, but to the uncertainties of his apparatus. Even so, there remains the outstanding problem—why so few neutrinos?

Neutrinos from supernovae?

The Sun is a weak source of neutrinos compared with a supernova.

Even before it forms a neutron star, the precursor of a supernova creates neutrinos in abundance by two main methods. In the first method, radiation creates matter. It does so in a beautifully symmetric

158

way. To every kind of particle of matter there corresponds a kind of particle of antimatter, and, if it is energetic enough, radiation can produce matter–antimatter pairs of particles. A gamma ray produced at the center of a massive star is energetic enough to create an electron and its antiparticle, a positron. These can recombine and produce a pair of neutrinos. In the second method, an electron is captured by a proton in a nucleus with the emission of a neutrino, and the resulting neutron decays back to an electron, a proton and a neutrino. The neutrinos run off with a fraction of the energy, but the original nucleus still remains to suffer again this process of attrition. This is called the *Urca process* after a casino in Rio de Janeiro, where the customer loses little by little.

The energy carried away by neutrinos from the center of a massive star is the very cause of the supernova explosion itself. Energy transformed into speeding neutrinos is lost from the star virtually instantaneously, whereas energy transformed into radiation jostles its way out of the star and helps to support the star against its own gravitational pull. The more energy lost from the center of the star as neutrinos, the less support the released energy gives to the star. When support drops too far, collapse becomes inevitable.

Initially, about half of the particles in the center of the star are neutrons; the other half are protons. Both neutrons and protons swim in a sea of electrons. The implosion forces the protons to swallow the electrons and make neutrons: the star center becomes a neutron star. Each creation of a neutron liberates one neutrino, increasing the neutrino output still further.

Is there a possibility of detecting neutrinos from supernovae? There's a bare possibility that they may have already been detected. Davis has performed more than 30 runs in his underground neutrino observatory, and in Run 27, performed in late 1972, a significant number of neutrinos was detected. There has been speculation that they were the result of a flash of neutrinos from a distant supernova, whose light went unrecorded because it was absorbed by dust in our Galaxy. However, there was another apparent increase in the neutrino detection rate in 1975–76 and two supernovae in three years seems too many. Perhaps these flashes are rare cosmic ray events, or experimental glitches, rather than supernovae. Davis's difficult experiment is giving tantalizingly equivocal results.

Is the experiment worth pursuing? It gives the first possible glimpse into the interior of the Sun and supernovae and the failure to observe the expected number of neutrinos has revealed astronomers' ignorance. Since the amount of material at and above the surface of a star—the only material visible to astronomers before the neutrino experiment—is just a few million millionths of the mass of the star and astronomers' idea of the structure of the remainder is educated guesswork, it is not surprising that calculations of the expected number of neutrinos from the Sun were wrong. The deficiency has inspired new efforts to understand better the interior of the Sun and other stars, though the American astronomer E. E. Salpeter's comment is still true, that "at the present time, we neither have a positive identification of solar neutrinos nor the morbid satisfaction of predicting a scandal in stellar evolution theory." More than this, the stimulus which the experiment has given to neutrino astronomy has meant that the properties of the neutrino have been subjected to particularly close scrutiny. Its interactions are so rare that terrestrial experiments have difficulty in detecting neutrinos at all. But in a supernova explosion so many neutrinos are released and the star is so dense that sufficient neutrinos may be absorbed to have a significant effect on the explosion of the star. The study of supernovae—cataclysms on the grandest scale—is revealing properties of the neutrino—nature on the tiniest scale.

XI   *Creation of the elements: Man, the supernova remnant*

Act first, this Earth, a stage so gloomed with woe
You all but sicken at the shifting scenes.
And yet be patient. Our Playwright may show
In some fifth Act what this wild drama means.

Tennyson

In some of the most rugged mountainous country of New South Wales, along the Turon River and in its hinterland, are the ghost towns of the Australian gold rush of the 1870s. A few people are left in Hill End, but of the town of Tambaroora nothing remains except Golden Gully, a canyon dug by miners in their thousands. Tens of millions of tons of dirt were dug from here, washed and scrutinized for the glint of a metal: precious because both beautiful and rare, and made rare by the rarity of supernovae.

Abundance of the elements

Lured to inhospitable terrain because of the incredible richness of the strike, those miners who were luckier than most took from the goldfield a total of just 20 tons of gold, on average one gram for every ton of dirt dug over. Modern gold mines operate profitably when there is a worthwhile concentration of gold of about 20 grams (about an ounce), per ton, in contrast with the average concentration of gold in the surface of the Earth of approximately one thousandth of a gram per ton.

161

Golden Gully, Tambaroora, New South Wales. Millions of tons of dirt were dug by pick and shovel from this gully to collect gold, grain by grain. The rarity of gold is linked to its formation deep inside massive stars. Because supernovae, scattering their heavy elements throughout space, are rare, gold is rare too.

162

What of the concentration of gold averaged over the entire mass of the Earth? There is, of course, no direct evidence since the central regions of the Earth are inaccessible. However, it is a speculation commonly held by astronomers that meteorites, stones and rocks which have fallen from space to the Earth, represent the remains of a defunct planet as solid as the Earth or Moon, so that the abundance of the elements in meteorites may be like the abundances in the interior of the Earth. Gold is 100 to 200 times more abundant in meteorites than in the surface of the Earth, so is presumably similarly more abundant in the Earth's center.

It is not difficult to guess why this should be. The Earth's interior is hot, because of heating by radioactive materials, and partly perhaps because of its contraction under the force of gravity. In some ways it resembles an ore-smelting furnace, melting rocks so that the lighter slag rises to the surface to make the Earth's crust, while the metals such as iron, nickel and gold fall towards the Earth's center.

Determining the abundance of gold in stars directly is not possible. Only one clear spectral line in the Sun's spectrum caused by atoms of gold has been found, proving gold's existence there, but since the mechanism in the gold atom that causes the spectral line has not been studied well enough, no accurate estimate of the amount of gold required to form the line can be inferred. Its concentration in the Sun has been estimated by assuming that, since nickel and gold have similar properties, the ratio of their concentration in meteorites is the same as in the Sun. Because nickel gives rise to many well-studied spectral lines in the solar spectrum, its abundance can be easily measured.

Approximately one milligram in every ton of the Sun is gold.

Why is gold rare? Why is iron on the other hand relatively common? Why is it that in spite of the great diversity of astronomical objects whose composition has been studied—the Earth, meteorites, the Sun, most stars—the relative abundances of the elements in all these bodies are surprisingly similar, and the differences are readily explainable by some plausible guesses and the histories of the bodies?

Clearly there is some common astronomical explanation for the origin

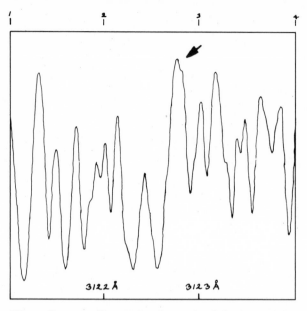

The valleys on this graph of 0.03% of the known spectrum of the Sun represent chunks of color "bitten" from sunlight by atoms, mostly of iron, in the Sun's atmosphere. Peaks represent colors which have passed relatively freely out of the Sun. The nibble arrowed is the only evidence for the existence of gold in them there solar hills. It hints at the wave of supernovae which manufactured the gold, and enriched the pre-solar nebula from which the Sun formed.

of the elements in all these celestial objects, and supernovae play a crucial role.

The alpha-beta-gamma theory

Modern discussions have a two-pronged attack on the creation of the elements. The first starts, as the Universe started, with the Big Bang. This theory of the origin of the elements was originally called the alpha-beta-gamma theory, in part because the elements are

164

supposed to be formed in sequence like the start of the Greek alphabet, and in part because the theory was proposed in detailed form by Ralph A. Alpher, Hans Bethe and George Gamow in 1948. (Bethe's part in creating the theory was small—Gamow said that he invited Bethe to be a coauthor because the pun appealed to Gamow's sense of humor.) Gamow called the material of the Big Bang *ylem*: he envisaged it as a gas made of neutrons, although modern authors see it as a more complicated mixture. When the theory was first put forward it was to account for the formation of all the elements from this basic ylem. As we shall see, however, it would complete only the first stage of the process.

The neutrons in ylem changed relatively slowly into protons and electrons as the Universe got under way. Some of the protons thus formed captured neutrons to make more complicated nuclei. Some of these nuclei would change by beta-decay (emission of an electron) and some would gather further neutrons to become more complex nuclei. All the element building in the Universe, according to the alpha-beta-gamma theory, occurred in the first two hours that the Universe existed. As the Universe cooled the nuclei would capture the free electrons to become atoms.

One feature of the abundance of the elements which this theory explains well is the fact that nuclei whose ability to capture neutrons is low are more common than those whose ability is high. Think of nuclei as a large intake of graduates into a big organization in which promotion is by merit. Bright graduates (nuclei with high neutron-capture ability) are susceptible to promotions; they take advantage of random opportunities (neutrons) as they occur and are promoted faster than their fellows (they form more complex nuclei). The duller graduates' ability is lower, they stay in their career grades longer and consequently there are more of them than of the bright graduates.

Among the elements certain nuclei have exceptionally low ability to capture neutrons—these occur at the so-called magic neutron numbers 50, 82 and 126. Elements occurring at these numbers, like lead, are exceptionally common.

165

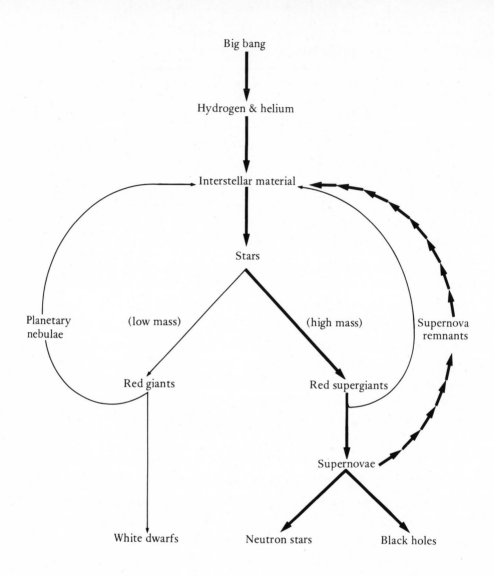

Big bang

Hydrogen & helium

Interstellar material

Stars

Planetary nebulae   (low mass)   (high mass)   Supernova remnants

Red giants   Red supergiants

Supernovae

White dwarfs   Neutron stars   Black holes

━ = route described in this book

Chemical evolution of the Galaxy. The flow is from the Big Bang which created hydrogen and helium from which stars form. Gas is recycled back into the interstellar medium via planetary nebulae or supernova remnants, or ejected by low surface gravity supergiants, but everything ultimately ends as white dwarf, neutron star or black hole.

The alpha-beta-gamma theory suffers, however, from a fatal flaw. It requires that the nuclei build up from hydrogen by adding one neutron at a time, with the neutron changing to a proton at appropriate stages. The flaw is that at two vital stages, the nuclei created cannot exist for more than a minuscule fraction of a second (a thousand, million, million, millionth of a second!) after which time they release the neutron that they have just captured and return to a helium nucleus again. They do not exist for long enough for the next step to occur. The break in the chain occurs just after the formation of helium, about two minutes after the start of the Big Bang. Thus the alpha-beta-gamma process cannot proceed past helium in making the elements. No further heavier elements can be made. If the alpha-beta-gamma process of element building were all that could occur, the Universe would consist solely of hydrogen and helium and there would be no carbon, no nitrogen or oxygen, no paper to make this book, no writer to write it, no reader to read it.

Creating elements inside stars

There must have been other sites in the Universe, apart from its beginning, at which the heavier elements were created. When the process by which nuclear energy became starlight was discovered as a result of the work by Bethe and von Weizsäcker in 1939, it was realized that the same processes would change the composition of the stars and create new elements.

The first observational evidence that the stars create elements was Paul Merrill's discovery in 1952 of spectral lines of the element technetium in red giant stars. Technetium is unstable and lasts at most for a few million years. As the red giant stars were known to be older than this, clearly technetium could not have been in these stars since they were formed; it must have been made there. Other stars were discovered which, like red giants, had developed sufficiently in their evolutionary life to show an excess of carbon or nitrogen caused by helium burning in the so-called triple-alpha process.

167

The time was ripe for a detailed examination of the formation of the elements in stars.

The problem engaged the attention of Fred Hoyle in 1946 because of his advocacy of the Steady State Theory of the Universe in which there was no Big Bang and therefore no cosmological element creation. Hoyle was faced with the existence of the wide variety of elements which he had to explain in other ways, and one of the main successes of the Steady State Theory was to stimulate this research, although the theory seems to have since lost credibility as a cosmology. The specific processes which form the elements in stars were detailed in a foundation-laying paper in 1957 by Geoffrey Burbidge, Margaret Burbidge, Willy Fowler and Fred Hoyle, known as the B^2FH (B-squared, F, H) paper. B^2FH supposed that the first stars consisted principally of hydrogen. Most stars visible now are in the process of converting that hydrogen to helium, releasing energy which can be seen as starlight.

This process creates helium from hydrogen. As stars age, they "burn" some of the helium which they have created to produce carbon, oxygen, neon and magnesium. The interior of the star may be mixed, stirred up by clouds of hot material billowing from the star's center towards its surface by the force of convection. Depending on whether mixing occurs or not, the carbon and oxygen can capture protons (hydrogen nuclei) or alpha particles (helium nuclei) to make either proton-rich nuclei or the elements magnesium, silicon, sulfur, argon and calcium.

B^2FH identified five further processes occurring in or on the stars. The first, which they named the e-process, occurs somewhere in some kind of star yet to be satisfactorily identified. When the mixture of elements formed in the previous processes cooks at a high temperature, the protons in the mixture absorb electrons and release them at equal rates in a situation of equilibrium (hence the name e-process). This creates those abundant elements such as iron, nickel, chromium and cobalt which are known collectively as the iron peak.

Up to this point the element building process has been relatively easy:

the creation of heavier elements has released energy. In a sense, the nuclei of these elements *want* to be formed (in the same sense that a heated object *wants* to cool by radiating energy). Beyond the iron peak however, the processes creating heavier elements have to have a supply of energy available to do so. Such a supply is available in a supernova explosion. Just before the supernova, the precursor star has built up a supply of middleweight elements by burning helium and carbon. Something happens and the precursor becomes unstable: the precise way in which the supernova occurs may not be important from the point of view of the creation of heavy elements. What matters is the speed at which the process occurs, so that neutrons are added to the middleweight nuclei sufficiently quickly that the successive nuclei, though unstable, do not have time to eject electrons and turn into something else. Because it is *r*apid, this process is called the *r*-process. (The *s*-process is one in which successive neutrons are *s*lowly added to middleweight nuclei which do have time to decay before the next neutron comes along.)

The *r*-process forms most of each of the following elements: selenium, bromine, krypton, rubidium, tellurium, iodine, xenon, europium, gadolinium, terbium, dysprosium, holmium, erbium, thulium, ytterbium, lutecium, rhenium, osmium, iridium, platinum, gold and uranium. The reader may find the names of some of these elements unfamiliar. Others he will know are precious because they are uncommon. These are consequences of the infrequency of supernovae.

A process called the *p*-process takes place when protons are added to nuclei formed by the *r* and *s* processes, but the *p*-process creates only a minority of the atoms of any one element. It may occur as the supernova ejects its outer layers into space, these layers containing unburned hydrogen and hence abundant protons, as well as *r* and *s* process nuclei.

Scattering the elements

The elements were made in stars. How do they come to be found in the Earth?

The elements created in stars, including those created in supernovae,

169

are thrown into interstellar space by supernovae themselves. The outer part of a supernova—perhaps most of it—is ejected into space by the force of the supernova explosion; this contains about one Earth mass of r-process elements. For a while—perhaps up to 100,000 years—the supernova remnant is visible in optical or radio telescopes but it soon merges with the general interstellar material, mixed up with the rest. New stars condense from the interstellar material, shining first as large cool stars, not visible to optical astronomers but emitting enough infra-red radiation to be detected by specialized infrared detecting telescopes. Such stars spin fast. Having contracted from a much larger slowly rotating interstellar cloud, new stars spin up as they contract, just as a neutron star spins up as it is exploded from the center of a supernova.

Nonetheless, because it is an observed fact that most stars spin slowly, newly formed stars must shed their angular speed by creating planetary systems such as ours. Virtually the whole of the rotational energy of the solar system lies in the massive planet Jupiter, which orbits the Sun once every 12 years. But if Jupiter, and the other planets, had not been ejected from the Sun, its rotational period would have been less than about three hours instead of its present period of just over 25 days. The newly formed Sun therefore shed some of its material in what is known as the solar nebula to make the solar system; and that is how the r-process elements, like gold, traveled from a supernova in the distant past and came to be found in the Earth, to be sought and dug up by the miners of the gold rushes.

Cosmological clocks

How distant a past was it? The study of the age of the elements is somewhat grandly called nucleocosmochronology, and it exploits the fact that some of the atomic nuclei created in the r-process are not stable—they spontaneously decay into something else, perhaps on a very long timescale indeed. For example, the nucleus of the iodine atom exists in two long-lived forms, one of which is very stable and lasts indefinitely, while the other decays into a form of xenon gas on a time scale of 17 million years. Both kinds of iodine are produced in approximately equal amounts by the r-process in supernovae, but because one, so-called iodine-129 changes to xenon-129, the other, iodine-127 eventually predominates.

While all the r-process elements are floating about in some interstellar cloud, the xenon-129 produced by the iodine-129 dissipates into space. But after the r-process elements have condensed into planets and meteorites, the xenon-129 produced by the decay of iodine-129 is trapped in the rock containing the iodine. The amount of xenon-129 in the rock thus tells how much iodine-129 has decayed since the rock solidified.

We know that there were equal amounts of iodine-127 and iodine-129 when the r-process which formed them took place; we know how much iodine-129 has converted into xenon-129 while the elements were in interstellar space. Because we know how long this takes, we now know the time that the r-process elements were in space before the rock solidified: some 200 to 600 million years (say half a billion years, in round numbers).

We have measured the time between the supernovae which formed the r-process elements which condensed into the solar system, and the formation of the solar system itself.

Is this a reasonable time? Stars form in the spiral arms of the Galaxy. There the dust and gas between existing stars is compressed enabling new stars to condense from it. Spiral arms swirl around the Galaxy, causing gas at a given place in the Galaxy to be compressed repeatedly, the time interval between compressions being measured in hundreds of millions of years, the rotation period of the Galaxy. If the elements dispersed into the interstellar medium by supernovae occurring between successive compressions are all condensed out to stars at each compression, the average time these elements are free in space would be roughly half the interval between compressions. If, as is more likely, just a fraction is condensed at each compression, the average time the elements are free in space would be longer, but would presumably still be measured in hundreds of millions of years—just about the length of time actually determined by nucleocosmochronology.

Nucleocosmochronologists (if, indeed, there are any astronomers who put up with the inconvenience of short gaps for "occupation" on their passports and call themselves this) can also estimate the total time since the formation of the r-process elements by looking at how much of two kinds of uranium exists now. Uranium-235 decays on a time scale of 710 million years whereas uranium-238 lasts much longer, 4.5 billion years. Both are formed from r-process elements in roughly equal amounts. When measured in rocks now uranium-238 predominates 138 times over uranium-235. It can be calculated that approximately 6.5 billion years must have elapsed since both kinds of uranium were formed.

The gold you own as a wedding ring, watch or tooth filling is thus the seed of some supernova which occurred some six and a half billion years ago, which drifted in space for up to half a billion years, which condensed as part of a protosun but was rejected from the body of the Sun, and became part of the solar nebula which condensed into the planets, including the Earth, and which found itself by chance geological processes near enough to the Earth's surface to be extracted by gold miners.

Metal-rich, metal-poor

Although the vast majority of stars which we see now contain the different elements in relative proportions which closely match the proportions in the Sun, astronomers have found stars which are deficient in the total amount of elements other than the hydrogen and helium produced at the origin of the Universe.

Thus, although, say, the ratio of the amount of iron to nickel is the same in most stars, the total amount of iron is highly variable from star to star.

Ignoring chemistry, astronomers have traditionally called all elements

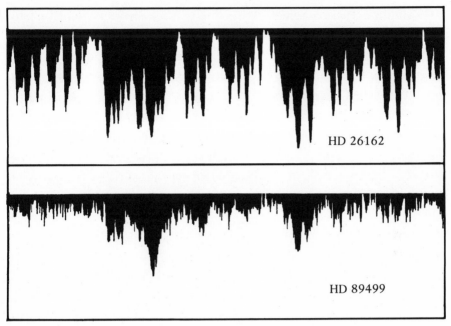

*In these graphs of the spectra of two stars the black areas (valleys) represent
colors lost from the star's output of light because of absorption by metal atoms in
its atmosphere. The bottom spectrum is of HD 89499, a star with a deficiency of
metals and therefore showing relatively little loss of light due to absorption. A
heavier blanket of absorption lies across the top spectrum, which is of star HD
26162 with a high metal content similar to that of the Sun. Olin Eggen of Mt.
Stromlo Observatory first drew attention to HD 89499, a star formed during the
initial collapse of our Galaxy, perhaps 8 billion years ago. Mike Bessell estimates
that it contains only 1% of the concentration of metals in the Sun.*

   *This diagram is based on spectra taken by Bob Fosbury and Mike Penston
with the 150-inch Anglo-Australian Telescope.*

other than hydrogen and helium "the metals," and stars with less than
normal amounts of iron and so on are termed *metal-poor*. The lines of
iron and other metals in their spectra are weaker than in other comparable stars signifying that there is a smaller quantity of metals present
to produce the spectral lines. Conversely *metal-rich* stars have stronger
than normal metal lines.

From the scenario of element creation that has already been sketched, the reader can guess that the older stars are metal-poor, having formed early in the history of the Galaxy before there had been many supernovae to make metals, while the very youngest stars are metal-rich, having formed recently from interstellar material which has been enriched by the detritus of supernovae throughout the whole of galactic history. The Hyades star cluster, that prominent V-shape in the constellation Taurus the Bull, is an example of a metal-rich cluster of stars having twice the concentration of metals of the Sun and an age of just half a billion years, compared with ages in excess of 10 billion years for the oldest metal-poor stars.

There are, however, no stars found having no metals at all, so the amount of metals in the Galaxy appears to have increased steadily from a non-zero initial value. In fact, the number of metal-poor stars in general is embarrassingly low for simple scenarios.

Where did the initial metals come from if they were not created in the Big Bang? Where have the metal-poor stars gone? Or were so few formed that very few still exist?

One possible answer to these questions is that early in the formation of our Galaxy there may have been a wave of star formation which preferentially produced large numbers of massive stars. Since massive stars quickly turn into supernovae, there would have been a burst of metal formation, and the wave of stars would soon have disappeared by becoming invisible black holes or faint neutron stars, pulsars whose ticking has long since faded to silence. This would imply that in the early years of the Galaxy there was a supernova every few days rather than every few decades as now.

With this answer it is difficult to maintain a balance between making this first generation of stars massive enough to do a convenient quick disappearing act, but not making them so massive that they overproduce metals which would still be around now in embarrassing overabundance. Among the solutions proposed to cope with this problem is a guess that primordial material left over from the formation of the Universe has been, and presumably still is, falling into our Galaxy to dilute the metals in the interstellar medium. No evidence for the existence of this infalling gas is known. Indeed, if there were a large number of supernovae during, say, the first two billion years of the existence of the Galaxy, significant amounts of gas would be blown out of our Galaxy, rather than falling into it.

You, the supernova remnant

In this subject a large theoretical superstructure has been erected on a narrow keel of observational fact and it cannot be said that the so-called Ultimate Model of the chemical history of our Galaxy has yet been conceived, let alone detailed. (It is, apparently, called the Ultimate Model so that when asked what it is, astronomers can reply "Um ....") But, according to Beatrice Tinsley in 1974, the scenario may proceed in five acts:

Act I. A Big Bang produces hydrogen and helium, in the early dense hot stages at the beginning of the Universe.

Act II. The Universe cools, becomes less dense and the lumps in it start to form galaxies. Actually, the lumps form galaxies more easily in the denser early stages than this later one, but nucleocosmochronologists turn a blind eye, a deaf ear and a cold shoulder to this difficulty.

Act III. Galaxies form into spiral or ellipticals depending on how they rotate. A portion of the gas in our Galaxy disappears during the first two billion years, forming new faint stars of low metal content, and the remainder collapses to a flat disk whose metal abundance is by then already half that of the solar system because of the first wave of supernovae.

Act IV. Our Galaxy continues to produce stars at a slowly decreasing rate for the next billion years (in the middle of which it produces the Sun). The rate of star formation drops slowly because gas is continually being locked up into neutron stars, white dwarfs and black holes, but the metal content of the gas remaining is continually increasing because each supernova enriches the gas with more metals.

175

Act V. The Sun having formed, a portion of the metals trapped in the rejected part of the solar nebula gives rise to the Earth, biological chemistry, evolution and the animal kingdom, including astronomers and readers of books. Looked at like this, people are the most interesting supernova remnants of all.

TABLE V   *Creation of the elements*

Process	Where?	Some of the elements formed
Alpha-beta-gamma process	Origin of Universe (Big Bang)	Hydrogen, helium
Hydrogen burning	Most stars	Helium
Helium burning (triple-alpha process)	Red giant stars	Carbon, oxygen
Carbon burning	Massive red supergiants	Neon, sodium, magnesium, silicon
e-process	{ Hot centers of stars? Supernovae?	Iron, nickel, chromium, cobalt
s-process	Evolved stars	Copper, zinc, lead, technetium
p-process	Surfaces or shells of supernovae	Small quantities of molybdenum samarium
r-process	Supernovae	Gold, platinum, "rare earths"
x-process	{ Surfaces of stars? Cosmic rays?	Lithium, beryllium, boron

176

XII *Cosmic rays: supernovae and evolution*

The crackle you hear and feel when you take off a sweater of synthetic fiber on a dry day is caused by static electricity. In the act of sliding the sweater over your head you rub its surface and fracture the atoms there, ionizing some and thus splitting off free electrons from the broken atoms. Some of the free electrons can be transferred from the sweater to you. You become charged with negative electricity, the sweater with corresponding positive electricity. If conditions are right your hair stands on end. The spare electrons from the sweater are feeling one another's repulsion, pushing their fellows as far away as possible, separating your hairs as much as they are able and making them stand on end like a porcupine's quills, or Hamlet's hairs when hearing his father's ghost's tale of purgatory.

You may touch a metal light switch and the electrons on you will flee into the switch and into the Earth, scattering away from each other as far as they can. But suppose you stand still; suppose you are wearing rubber-soled shoes through which the electrons cannot pass. As you watch in the mirror you see your hair gradually settle into place as the excess of electrons in your body gradually leaves it. In a matter of minutes they have dissipated.

How did the electrons on your body dissipate into the air? Dry air is a very good insulator, like rubber—electrons do not pass easily in air.

What conducting threads reach through air from you to the Earth, along which electrons may pass?

Attempts to understand this phenomenon started in 1900. The electron had been discovered three years earlier by J. J. Thomson and its role in the phenomenon of static electricity was being investigated by Elster and Geitel in Germany and C. T. R. Wilson in England. They established that even dry, pure air was not made up solely of complete atoms but that some of its atoms had broken into fragments consisting of electrons, and what was left, ions. Discharge of static electricity takes place by electrons skipping from ion to ion to reach the ground. But what causes the ionization of air?

The early experimenters attempted to find how the degree of ionization of air varied with atmospheric conditions, geographical location and time of day. For this purpose they developed a carefully insulated device known as an electroscope which would retain static electricity well except for the electrons that it discharged through the air. By watching its rate of discharge they hoped to obtain a clue to the amount of ionization of the surrounding air.

Suspecting that ionization of air was caused by radioactive rocks such as uranium-bearing ores, Wulf and Gockel took an electroscope to lakes and glaciers which, being of relatively pure water and ice, generated little radioactivity. The ionization fell considerably. Some ionization of air must therefore be caused by radioactivity. There was however a degree of ionization even over the thickest glacier, the deepest freshwater lake. Was there some residual radioactivity even in fresh water? Did moving the electroscope higher above the glaciers or lakes diminish the residual effect? Curiously it did not.

If natural radioactivity from rocks was the cause of the ionization, did this effect diminish with altitude? Wulf took his electroscope to the top of the Eiffel Tower in Paris (330 meters high). There was some decline but not enough. V. F. Hess in Vienna decided to investigate the effect of altitude by ascending in a balloon. His first flights to 1000 m showed a small reduction of ionization in the air. By 1912 he had made ascents to more than 5 km and, to his astonishment, he found that the ionization actually began to *increase* at heights above 2 km. He concluded that radiation must be penetrating the atmosphere *from above*. The obvious source of such radiation was the Sun, but this had to be ruled out when Hess made an ascent during a partial eclipse. Although the Moon had obscured part of the Sun, there was no drop in the ionization of the air. Similarly, there was no difference between daytime and nighttime levels. Nor could the 23 hour 56 minute period of the Earth's spin beneath the stars be picked out—so whatever space radiation was ionizing the air could not come from any one object.

Explaining cosmic rays

These mysterious emanations were first termed *cosmic radiation* by the great American physicist R. Millikan, who began research in the subject after World War I. He also realized that cosmic rays were bringing an energy comparable to that of starlight to the Earth.

Gradually the complex properties of the cosmic rays became understood. The modern picture is that the cosmic rays incident on the Earth from space are atoms, stripped of most or all of their electrons and therefore are bare nuclei, moving at speeds close to that of light.

Cosmic rays are influenced by the magnetic forces in the Galaxy and the solar system. The lowest energy cosmic rays are deflected by magnetized clouds ejected from the Sun during periods of solar activity. The Sun's activity varies with an 11 year cycle. During its most active phase there are many large sunspots on its surface and many solar "storms" which eject the clouds into the space between the planets. These clouds cocoon the Earth and shield it from the influence of interstellar cosmic rays, though the Sun itself may contribute some extra cosmic rays to the flux striking Earth. The magnetic field of the Earth itself also plays a part in deflecting cosmic rays.

Those cosmic rays which do reach Earth's upper atmosphere collide with air atoms to make showers of electrons and other particles of matter which cascade down to the Earth's surface. These secondary particles are the ones which ionize air near sea level and cause the gradual discharge of static electricity from electroscope or tousled hair.

Much of the early study was of the secondary cosmic rays; only as balloons and rockets reached far above Earth did the properties of the primary cosmic rays become clearer. Present-day studies have used cosmic ray detectors in space, both on artificial satellites and on the Earth's natural satellite, the Moon. The latter experiments were set up on the Moon's surface during the Apollo 16 lunar flight and brought back to Earth afterwards.

The abundances of the different nuclei found among the cosmic rays follow closely the abundances of the nuclei determined from the cosmo-chemistry of meteorites and the like as discussed in the last chapter—hydrogen is the most abundant, and the 90-odd remaining heavier elements are less and less common as they increase in complexity. There are three major exceptions—the elements just heavier than helium, namely lithium, beryllium and boron, are a million times more abundant in the cosmic rays than in other matter in the Universe.

We saw in the last chapter that the creation of elements in the Big Bang could not get to lithium. Moreover, it turns out that nuclei of all three of these elements are readily destroyed at temperatures of about a million degrees—relatively low compared with those found in stars. Far from lithium, beryllium and boron being created in stars, therefore, any interstellar material condensing into a star and then being returned to space in a supernova or whatever has been purged of these elements, unless they have been made on the relatively cooler stellar surface or in space itself. If these three elements can be made neither in the Big Bang nor in stars, where can their origin lie? The B^2FH paper of the Burbidges, Fowler and Hoyle named the process which created these elements the x-process because it was unknown.

The x-process may in fact be the travel of cosmic rays through the interstellar material. The cosmic rays contain carbon, nitrogen and oxygen nuclei in relatively high abundance. As such nuclei collide with hydrogen and helium in interstellar space, the individual protons and neutrons in the nuclei which collide are rearranged into several fragments of different sizes including lithium, beryllium and boron nuclei. (This process is also called *spallation*.) The cosmic rays might therefore be expected to contain an excess of spallation products. They contain not only the great excess of lithium, beryllium and boron, but also other spallation elements such as chlorine and manganese.

From the amount of these elements in cosmic rays, cosmic ray physicists can establish through how much interstellar matter the average cosmic ray has traveled so as to produce the observed quantity of spallation products. The average cosmic ray has been traveling in the Galaxy for some 10 to 100 million years. After this time it leaks from the Galaxy into intergalactic space. Though a long time on a human time scale, 10 to 100 million years is short when compared to the age of the Universe (measured in units of 10 billion years) and the lifetime of most stars. If it is to the stars that we must look for the creation of cosmic rays, it is to the short lived, massive ones that we must direct our attention, and these are the ones which turn into supernovae. In fact Baade and Zwicky calculated that supernovae occur often enough and have enough energy to account for the supply of all the cosmic rays in the Galaxy.

The way in which they do so is not clear. Some believe that cosmic rays are injected into space during the supernova explosion itself. Others say that it is not the supernova which supplies cosmic rays to the Galaxy but the rotating neutron star or pulsar, which the supernova forms. This sweeps up interstellar material and flings it into the Galaxy at great speed.

Perhaps a mixture of these processes is responsible for pumping cosmic rays into space to stream about the Galaxy, and, before they escape from it, possibly to collide with Earth.

The influence of cosmic rays

Do the cosmic rays have more profound effects on the Earth than influencing the discharge of static electricity? In fact it appears that they do. Because of supernovae, archaeologists have a time scale for dating in prehistory.

The spallation reactions occurring as the cosmic rays collide with Earth's atmosphere produce neutrons, most of which are eventually absorbed by the predominant gas in the atmosphere, nitrogen. In this way the nitrogen turns into a carbon atom, of a kind called carbon-14 having an excess of two neutrons over the more usual carbon-12. Because of this, carbon-14 is unstable and decays to nitrogen-14, half of it changing every 5568 years. Though the carbon-14 is produced high in the atmosphere, at an altitude of 10 to 15 km, it combines with oxygen to make carbon dioxide, and diffuses rapidly to Earth's surface.

Carbon dioxide is a gas which is abundant in the atmosphere and which is "breathed in" by plants to take part in their cell building processes. In this way, therefore a small proportion of the carbon atoms which the plants accumulate during their lifetimes are radioactive carbon-14 atoms. Because plants are eaten by animals, all living organic material contains this proportion of radioactive carbon. When the plant or animal dies, the assimilation of carbon stops and the carbon-14 begins its long, steady decay to ordinary carbon. If the organism has recently died, its cells will contain their full abundance of carbon-14; if it has been dead a long time, the proportion of carbon-14 will be small. The ratio of ordinary carbon to carbon-14 is therefore a clock which runs down at a known rate.

In the cases where parts of the dead organic matter remain preserved—as timbers in a house, or bones in a midden, for example—this clock can be used to measure with some accuracy the year in which the tree was felled or the animal died. This technique, known as radiocarbon dating, can be used to estimate the age of material up to 20,000 years old, and is therefore of inestimable value in archaeology.

Now this method relies on the level of cosmic radiation having been constant for many thousands of years. Suppose there is a sudden significant burst of cosmic radiation from space, producing more carbon-14 than normal. When this carbon-14 is assimilated by living organisms, their radioactivity level is suddenly much higher. After death, the organic material still exhibits a proportionally higher radioactivity level. It would seem to the archeologist that the clock was still tightly wound, and that the organism was more recently dead than was the case.

There is good evidence that the production of carbon-14 has indeed been variable in the past. Evidence for the variation comes from using two different methods to estimate the age of wood from the incredibly long-lived California bristlecone pine.

One method is radiocarbon dating, while the other is called dendro-chronology—counting tree rings. Each year, a tree adds a new ring to its circumference. Counting these rings offers a potentially accurate way of determining the age of a tree whose date of death or felling is known, and bristlecone pines as old as 4600 years have been found. By overlapping ring sequences from many bristlecone pines, C. W. Ferguson has established the age of wood which formed 8000 years ago. Carbon-14 dating of wood as old as that, however, seems to suggest that it is some 900 years younger. Only over the last 2500 years do the two dating methods agree on average. Evidently the rate of production of carbon-14 was once higher than now.

Was this the result of a sudden burst of supernova activity increasing the number of cosmic rays in the Galaxy? The Czech geophysicist V. Bucha offers another explanation: that it was due to changes in the Earth's magnetism and its corresponding ability to shield the atmosphere from cosmic rays.

There is no real evidence for a significant change in the cosmic ray flux in historical times. The record of "fossil" cosmic rays has been extended back to 10 million years by a count of the number of tracks left by cosmic rays as they struck meteorites orbiting the solar system, showing no major changes in that time interval. Before then, the cosmic

183

ray intensity in the Galaxy could have been different from now, and presumably, was much larger very long ago during the first wave of formation of massive stars in the Galaxy and the consequent high rate of supernovae. But we have no direct evidence about the cosmic ray rate at this time.

Did a supernova kill the dinosaurs?

In 1957 Iosif Shklovsky considered what would occur if a supernova exploded near to the Sun, within say ten parsecs. The supernova would shine at magnitude -20 for a matter of months (not as bright as the Sun but far brighter than the Full Moon). After a few thousand years, the gaseous envelope of the supernova, ejected from it, would pass across the solar system. If the supernova were like the Crab Nebula, the synchrotron radiation trapped within the filaments would be seen as bright as the Milky Way, filling half the night sky. There would be no dynamical changes in the solar system—the amount of mass striking the planets would be too small to deflect them from their orbits at all. However, the solar system would be within the supernova shell for some 10,000 years, and the density of cosmic rays would be 10 to 100 times its current value. There would be much more cosmic radiation striking the Earth's atmosphere and much more radiation at sea level caused by the secondary cosmic rays.

Shklovsky has pointed out that such an increased level of background radioactivity might have played a part in the evolution of life on Earth. Evolution proceeds as a result of chance mutations in individual biological organisms. Some mutations better fit the organism to survive and, during its longer lifetime, it has a better chance to perpetuate the mutation to future generations. Most mutations, however, are unfavorable, particularly when the change to the genetic material has been gross. It might be expected that a large increase in radioactivity would produce many gross changes in genes and thus tend to cause organisms to die and become extinct.

Two thirds of the mean radioactivity on Earth is caused by terrestrial factors, mostly natural radioactivity in rocks at Earth's surface. One third is caused by cosmic rays. If the cosmic ray intensity increased 100-fold, the background radioactivity would be some 30 times increased. Shklovsky speculates that the dying out of the giant reptiles at the end of the Cretaceous period was the consequence of such an increase, caused by the sudden bathing of the Earth in the ejecta of a close supernova unwitnessed save by the uncomprehending eyes of doomed prehistoric dinosaurs.

This speculation about the reason why dinosaurs are extinct is one of a class termed "cataclysmic" by paleontologists—the class includes all global catastrophes which might exterminate a widespread order of animals. But there is no evidence in the geological record for the extermination of other orders at the same time. Dinosaurs did not live alone and their contemporaries survived them. Paleontologists generally therefore accept no cataclysmic reason for the extinction of the dinosaurs, and Shklovsky's attractive notion is not widely accepted.

In an interesting development of the theory that cosmic rays cause mutations, George Michanowsky proposed that cosmic rays from the Vela supernova, occurring about 8000 BC, triggered off a new awareness in men's minds and precipitated the technological era which we are still witnessing. If this is right the Vela supernova would be like the intelligence-donating obelisk in Arthur C. Clarke's book and movie, *2001 — A Space Odyssey*, and might have had a physical effect on mankind comparable to the mental liberation sparked by Tycho's supernova in 1572.

This is just speculation. Nonetheless it must be true that cosmic rays, by working on the genes of all species, inevitably cause mutations. Since, in fact, the natural radioactivity in rocks is a consequence of the formation of radioactive heavy elements in supernova explosions, and since the cosmic rays are generated in supernovae explosions, it could be said that, having played a crucial part in producing and distributing the chemical elements which make life possible at all, the supernovae are responsible for life's further evolution.

185

XIII   *Supernovae in binary stars*

Only about 15 per cent of all stars are in what we might call the "lone star state"—that is, they are single, like the Sun. Nearly half (46 per cent) have partners, with one star orbiting the other; the remainder (39 per cent) occur in multiple star systems.

It must therefore be common for a supernova to occur in a multiple star system.

The star which explodes first in a double or binary star system is the larger of the two since more massive stars expend their nuclear energy at a faster rate than smaller ones, and pass more rapidly to the late stages of stellar evolution.

What happens when the more massive star of a pair explodes as a supernova? Certainly the supernova explosion has some effect. In the first place, the smaller companion star receives the impact of the explosion and recoils like a catcher's mitt grasping a fast-thrown ball. If the companion star was describing a circular orbit about the exploding star, the orbit is now eccentric or flattened to some degree. The companion may also be blasted by stellar shrapnel, and material from the exploding star may significantly increase the mass of its companion. Even if it was heading to a quiet end as a white dwarf before the supernova explosion, it may now itself be too large for this and be destined to be a supernova too. In this sense, at least, supernovae can cause supernovae.

Not all the exploding material accretes onto the companion star. Perhaps most is ejected into space. It is even possible that the blast from a supernova is so severe that it strips the outer layers from its companion, decreasing its mass and lengthening its life. All in all, more than half the total mass of the double star system can be blown off into space. In this case the pair is split asunder, each star flying away like a stone flung from a slingshot. As a numerical example, consider a 30 solar mass star and a 4 solar mass star orbiting each other, making a total of 34 solar masses. The large star explodes leaving a 2 solar mass neutron star behind and ejecting 28 solar masses. Eight solar masses of this may fall onto the smaller companion, now making it 12 solar masses, while 20 solar masses of material is ejected into space. Because this is more than half the original mass of the double star, the neutron star and the 12 solar mass star are flung in opposite directions, speeding away from each other. Where there was once a double star, now there is simply an empty space at the center of an expanding shell of fragments, with two stars rushing away.

Where might we find such events? Stars of 30 solar masses are profligate with their nuclear energy and have lifetimes measured in only a few millions of years, compared with the 10 billion-year lifetime of the Galaxy. They are thus young objects and must have been recently formed from the interstellar gas in the Galaxy. This gas only occurs in the plane of the Galaxy in spiral arms. Thus bright, massive stars are found near the galactic plane.

If one star of a pair goes supernova and disrupts the double star, the two stars are quite likely to be flung out of the galactic plane altogether. This ties in well with the observations of radio pulsars, which are indeed found high above the galactic plane. The nine whose proper motions have been measured have an average speed in excess of 250 km/s, compared with a typical average speed of 15 km/s for stars just formed from the interstellar material. The evidence thus suggests that radio pulsars are formed from a binary star system during a supernova explosion which disrupts the binary star.

187

This is confirmed by the fact that, in contrast to the average population of stars where a majority are in multiple star systems, only one of the 149 pulsars known is in a double star. Even this lone example of a pulsar in a binary star is describing a very eccentric and elongated orbit about its stellar companion, suggesting that it only just failed to be severed from the double star system.

What of the former companions of the pulsars? Among the normal slow-moving population of bright, massive stars in our Galaxy there are some so-called "runaway" stars, with high speeds often exceeding 100 km/s. Three of these are Mu Columbae, AE Aurigae and 53 Arietis. From their speeds and directions, astronomers calculate that each left the region of the constellation of Orion some three million years ago, though they are now in the widely separated constellations of Columba, Auriga and Aries. Orion contains a large number of massive stars, and it was postulated by Adriaan Blaauw in 1961, following a suggestion by Fritz Zwicky, that these three stars were in a quadruple star system, the fourth member of which exploded as a supernova flinging the three runaway stars far from their birthplace.

Many astronomers believe this to be the general explanation for the massive runaway stars. It is also true that the runaway stars themselves will eventually undergo a supernova explosion, turning them into pulsars like their former companions, and they will then continue in their high speed flight beyond the plane of the Galaxy.

Some people have speculated that the Crab Nebula supernova was originally a runaway star formed by the earlier disruption of a binary star in the supernova explosion that produced another pulsar near to the Crab Nebula, namely NP 0527, which has the peculiarity of having the slowest pulse rate of any pulsar. The hypothesis was that both the Crab Nebula pulsar and NP 0527 were once members of a binary star system in the association of massive stars called I Geminorum. However, the problem with this attractive idea is that the space velocity of NP 0527 is too slow for it to have traveled from this association to where it now lies, within the short lifetime of a pulsar. The hypothesis is not well

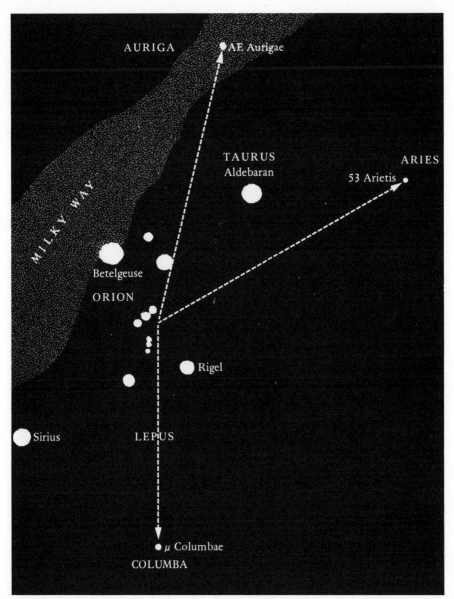

Runaway stars AE Aurigae, 53 Arietis and Mu Columbae were thrown like slingshots from the constellation Orion three million years ago when a fourth star of a quadruple star system disintegrated in a supernova explosion. Each has traveled right across an intervening constellation to reach the present-day positions 30 to 40 degrees from the origin.

substantiated. However the Crab Nebula pulsar itself does have a considerable proper motion away from the constellation Gemini and it may be that the supernova of 1054 occurred on a runaway star ejected about four million years ago from the I Geminorum association of massive stars.

It is not inconceivable that the momentum of the runaway stars and pulsars will in some cases be sufficient to carry them right out of the Galaxy. If the speed of the pulsars near the Sun exceeds some 290 km/s they have more than the velocity of escape from the Galaxy and will escape from it into intergalactic space. Pulsar number 0283 + 36 probably has such a speed and may be doomed to become an intergalactic tramp.

Presumably not every binary star system in which a supernova explosion takes place is disrupted. Sometimes the amount of material ejected will be less than half the total mass of the double star system so the laws of physics dictate that the pair will not break apart.

Perhaps the explosion itself will cause the pulsar formed to recoil in such a way that it remains in orbit around its companion. Why is it then that binary pulsars are so rare? The answer is presumably that the huge envelope of ionized gas from the ordinary star surrounds the pair and traps any detectable radio emission from the pulsar. Such a cloud of gas is blown off by our own Sun, and is known as the solar wind. However the pulsar does ultimately become detectable—as an X-ray pulsar.

Why a pulsar shines X-rays

In the course of time the hydrogen fuel in the center of the companion of an unseen pulsar in a binary star system gives out and the star begins to expand into a red giant (as outlined in Chapter IX). It may be that the star and pulsar are sufficiently close that before the star's growth to a red giant is complete its atmosphere begins to leak onto the pulsar.

It is easy to see that at a certain point between two stars their gravitational forces exactly cancel each other, so that an atom placed at this point might find it difficult to decide to which star it should fall. Jules Verne wrote a story in which the crew of a giant shell fired into space suddenly fell from the floor to the ceiling of their capsule as they passed the equivalent point between Earth and Moon. (He did not understand the concept of free fall, and that astronauts are weightless except when their space capsule is being propelled.) The real situation in a binary star is complicated by the presence of centrifugal forces caused by the revolution of the two stars in orbit around each other, but nonetheless the gravitational field of each star has its own zone of influence within which all material belongs to that star. These areas are teardrop shaped with the points touching, and are called Roche lobes. When one star fills its Roche lobe its atmosphere may protrude beyond its lobe into the adjacent one and fall towards the other star. So gas is transferred from the putative red giant to the pulsar.

Now, the pulsar is small: it is a neutron star, some 20 km in diameter. When the gas from its companion falls upon the neutron star, the gas is compressed by the intense gravitational force. Just as compressing the air in a bicycle pump heats it, so the infalling gas is heated, to a temperature which may be tens of millions of degrees. The hotter a body the more energetic (shorter wavelength) the radiation it emits. This gas is so hot that it shines not by emitting infrared radiation or light but by emitting X-rays. The neutron star becomes detectable as an X-ray star.

The power available from the gravitational field of a neutron star is enormous. A marshmallow falling onto the surface of a neutron star would explode with the energy release of a World War II atomic bomb.

Many X-ray binary stars are now known from observations by X-ray survey telescopes on board artificial satellites, particularly Uhuru, launched in 1970, and the Copernicus satellite launched in 1972.

How do astronomers know that a particular X-ray source is part of a binary star? Take as an example the strongest X-ray source in Hercules, called Her x-1. This source switches off for $6\frac{1}{2}$ hours every 1.7 days. Similar behavior is found among ordinary stars, too, the explanation being that they are double stars of the type called *eclipsing binaries*. Quite simply, one star hides the other for a while during its orbit because the plane of their orbits is roughly in our line of sight.

In the case of Her x-1, the companion was known to be a variable, called HZ Herculis, before the X-rays were detected coming from the source. It is variable because the X-rays shining from the neutron star heat one side, this side turning towards and away from Earth with the orbital period of the neutron star. Her x-1 itself is a pulsar, pulsing X-rays with a period of nearly one and a quarter seconds. As the pulsar orbits HZ Herculis, the Doppler shift of its pulsing frequency can be clearly detected. In fact, the Doppler shift of the companion star can also be measured, since some of the pulsed X-rays are intercepted by the companion star where they heat up its surface and are re-emitted as pulsing visible light.

A sustained and difficult monitoring of this weak pulsation has been made by Berkeley astronomers Jerry Nelson and John Middleditch so that they have been able to measure the Doppler shift of the light pulses emitted from the companion star and determine its orbit about the neutron star.

From these measurements, astronomers can tell at exactly how many kilometers a second each star is traveling. Combining this with the orbital period of 1.7 days and the fact that the orbits are in our line of sight, all details of the orbits are known. The importance of this is that the masses of the two objects can be worked out, using the same orbital laws that Kepler deduced from observations of the solar system.

At last astronomers have been able to get to grips with observational facts about a neutron star. Theoreticians had calculated that no neutron star could have a mass greater than twice that of the Sun. In vindication of their work, the mass of Her x-1 turned out to be 1.3 solar masses.

Unlike radio pulsars which are all slowing down, the two best studied X-ray pulsars, Her x-1 and Cen x-3, are speeding up. This is probably because X-ray pulsars are members of binary systems by their very nature. Impact of the infalling matter onto the surface of the neutron star gives an impulse to the star, speeding it up.

A nova flares

Astronomers now believe that processes in which close binary stars feed upon one another are quite common and are responsible for another type of stellar cataclysm—novae. In recent years plausible theories have been put forward which link novae with events in close binary systems consisting of a red giant and white dwarf.

We have seen that double stars are common, and have shown how material can transfer from one star to another when a red giant star expands outside its Roche lobe. When the material falls onto a pulsar it heats up and emits X-rays, creating a pulsating X-ray source. But when material falls onto the already hot surface of a white dwarf star, the material may build up to form a shell of hydrogen within which stellar nuclear reactions may take place. The surface then explodes, and a nova results.

The total mass thrown off the white dwarf is typically only a few millionths of the original mass of the binary star system. For all its drama, and unlike a supernova explosion, the nova phenomenon does not penetrate deeply into the white dwarf. Nonetheless the mass thrown off is sometimes detectable as it causes a change in the period of the binary star. The nova in the constellation Hercules in the year 1934 was known as an eclipsing binary star with a period of 0.1932084 day before its outburst. Afterwards its period had lengthened to 0.1936206 day.

Gas continues to flow from the red giant onto the white dwarf after a nova explosion and there can be a further buildup of hydrogen and repeated nova explosions. Stars in which more than one nova explosion has occurred are called recurrent novae; it may be, given a long enough time, that most novae are recurrent. A recurrent nova in the southern constellation Pyxis holds the record for the number of outbursts, with a total of 4 in the last 80 years.

Astronomers are pretty sure that they are on the right lines in talking about close binary systems as an explanation for both pulsating X-ray sources and optical novae. But almost monthly, X-ray satellites such as

Ariel 5 are picking up new kinds of flaring sources which astronomers term transient stars, and where these fit into the picture is not clear.

There are the *bursters*, which flare up within seconds and fade away quickly, so the whole event is over within a couple of minutes. Some sources flare and die regularly, with a large flare being followed by a long quiet period, while a small flare is followed by a shorter quiet period. It is very likely that such behavior is caused by a buildup of material in the flow from one star to another and its sudden heating.

Then there are the X-ray novae, which can dominate the X-ray sky for a few weeks at a time. Only two of these have, at the time of writing, been linked with optical novae. Probably what is seen is the side of the companion star facing the neutron star, heated by the outburst. As the X-ray nova dies away so the companion star cools and its burst of light fades too—the fading shadow of a brief candle.

TABLE VI   *X-ray binary stars*

X-ray name	Star name	Distance	Binary period	Pulsar period	Mass of star	Mass of companion	Nature of companion
		(l.y.)	(days)	(sec)	(units of Sun's mass)		
Her x-1	HZ Her	12,000	1.70	1.24	2	1.3	neutron star
Cen x-3	Krzeminski's star	21,000	2.09	4.84	18	1	neutron star
SMC x-1	Sk 160	210,000	3.89	0.716	28	about 2	neutron star
Cyg x-1	HDE 226868	7500	5.60	none	15	10?	black hole?

XIV *Black holes from supernovae*

The bigger they come, the harder they fall.
R. Fitzsimmons,
champion boxer

Not all supernovae produce neutron stars. Searches at the centers of supernova remnants formed by recent supernovae such as Tycho's, Kepler's and the Cas A supernovae remnants have revealed no pulsars. Perhaps the beams of these pulsars never point to Earth in their rotations. But there are too many supernova remnants devoid of pulsars for this to be the likely explanation for all. Perhaps supernovae produce other, stranger stars which we cannot see.

How to escape

A stone tossed in the air loses speed and momentarily halts at its highest point before plummeting back to Earth again. When hurled with greater force, the stone rises higher but is still drawn back by the force of gravity. There is clearly a relation between the initial speed given to the stone and the height to which it rises: scientifically speaking, that momentary halt at maximum height is where the stone's potential energy due to the force of gravity exactly equals the kinetic energy it was given as it was thrown. The question arises: what gravitational energy does the stone have at higher points above the Earth and what speed is it necessary to give the stone to throw it arbitrarily far above the Earth, to "infinity"? Is it possible for the stone to be hurled with sufficient speed for it to leave the Earth completely and escape its gravitational pull? It is, and this speed is called the *velocity of escape.* The velocity of escape from the Earth's surface is 25,000 miles an hour: an object thrown with this speed from Earth does not fall back.

How does the velocity of escape vary from place to place—from planet to planet, star to star? The larger an astronomical body, the less the force of gravity at its surface; but the more massive it is, the greater its gravitational force. Put in mathematical terms, the velocity of escape is proportional to the square root of the mass of the body divided by its radius. Jupiter has 318 times the mass of the Earth and a radius 11 times

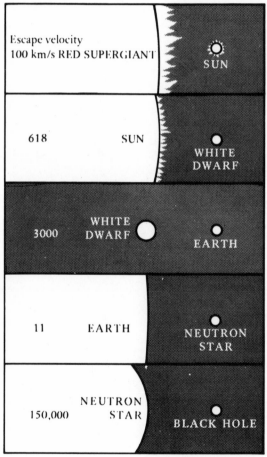

Escape velocity
100 km/s RED SUPERGIANT   SUN

618   SUN   WHITE DWARF

3000   WHITE DWARF   EARTH

11   EARTH   NEUTRON STAR

150,000   NEUTRON STAR   BLACK HOLE

Comparative sizes and escape velocity of the Sun, Earth and some stars. Escaping from Earth needs an 11 km/s impulse and is easiest because the Earth's mass is small. A red supergiant is the easiest star to escape from as its size is large. Escaping from other stars becomes more difficult as the stars become more compact. Escaping from the most compact objects known, black holes, is impossible, requiring a speed greater than that of light.

larger than the Earth's: the escape velocity from Jupiter is therefore the square root of 318/11, which is 5.4 times that from Earth.

The escape velocity from the Sun is more than a million miles an hour: though larger than the Earth it is much more massive and so has a larger velocity of escape. Curiously, although other dwarf stars like the Sun range in mass from a few hundredths of the Sun's mass to 60 times its mass, the range of sizes almost exactly compensates for this, with the result that the escape velocity from all other dwarf stars is not much different from that of the Sun.

Only when stars have finished the phase in which they burn hydrogen and expand to become red giants and supergiants does their escape velocity alter much from the solar value. In a red supergiant the escape velocity can be less than one sixth the solar value. This is the reason why matter can relatively easily escape from stars at the red giant and super-giant stage so that some massive stars can bring their mass under the Chandrasekhar limit and become white dwarfs in their subsequent evolution.

As you might expect, white dwarfs have high escape velocities. Even though their masses are similar to that of the Sun, they are denser and more compact, with radiuses typically one hundredth the Sun's. From the square-root law, therefore, the escape velocity is typically ten times as high as the Sun's. If matter is to escape from a white dwarf, such as in a nova explosion, when the white dwarf's outer envelope is thrown off, the explosion must be energetic enough to give the envelope an escape velocity of thousands of kilometers per second.

The concept of escape velocity is useful in describing not only the energy required to blast matter off the surface of a star, but also the loss of energy by radiation as it is emitted from the star. Radiation too loses energy as it travels against the force of gravity. This fact is not obvious from ordinary experience but is a feature of Einstein's General Theory of Relativity. Although very small in the Earth's gravitational field, the loss of energy has nonetheless been measured in gamma rays traveling up a mine shaft.

The lower the energy of radiation, the longer its wavelength. As light loses energy climbing out of a gravitational field, therefore, its color shifts towards the red end of the spectrum. This phenomenon is termed the *gravitational redshift*. The fraction of energy lost by radiation as it leaves the surface of a star is the square of the velocity of escape measured as a fraction of the speed of light. The gravitational redshift is not measurable in most stars, as we can see by taking the Sun as an example. Since the velocity of escape from the Sun is about 600 km/s and the speed of light is 300,000 km/s, the fraction of energy lost by light as it leaves the Sun's surface is only $(600/300,000)^2$ or one part in a quarter million.

But although the Sun's gravitational redshift is barely detectable, soon after white dwarfs had been discovered Arthur Stanley Eddington in 1924 pointed out they they had a large escape velocity and that measurable redshifts could be expected in their spectra. The escape velocity from a white dwarf is typically 3000 km/s (Table VIII), so the redshift is $(3000/300,000)^2$, or one part in 10,000. Thus H-alpha light emitted from hydrogen at a dense white dwarf's surface with wavelength 6563 angstrom units might be seen by an observer on Earth with wavelength almost 6564 angstroms, a small but measurable increase. There was, however, a confusing detail: there is no way of distinguishing a gravitational redshift from the more familiar Doppler redshift caused by possible motion of the white dwarf away from Earth, at a speed of a few tens of kilometers per second or so, unless the speed could be accurately known, and allowed for.

The only way round this problem was by finding white dwarfs which were part of binary systems, sharing a common motion with other ordinary stars. W. S. Adams measured the shift in the spectrum of the white dwarf Sirius B, which is in orbit around the bright star Sirius itself. The speed of Sirius is 8 km/s towards the Sun, the speed of motion of Sirius B around Sirius could be accurately calculated from knowledge of its orbit, and the redshift still unaccounted for was 21 km/s compared with the value of 20 km/s calculated by Eddington. The redshift of

40 Eridani B, a white dwarf in orbit around the star 40 Eridani, has been similarly measured by D. M. Popper, with good agreement with the theoretical value.

Measuring the gravitational redshift from the surface of a neutron star such as the Crab Nebula pulsar would be very interesting, since the wavelength of light emitted from its surface would be much changed from its original value. Formerly invisible ultraviolet light would be redshifted into the visible part of the spectrum, and the shift would give astronomers a way to estimate the mass and radius of a pulsar. In fact there are no atoms at the surface of a neutron star to emit light of a distinct wavelength: the redshift cannot be measured, although it would be very large.

Possibly enormous gravitational redshifts may be responsible for the behavior of high redshift quasars, mysterious starlike objects lying among other galaxies. Some, perhaps most, astronomers argue however that the redshift of quasars is caused by their motion in the Universe as they participate in the explosion of the Big Bang and recede from us, and that the high values of the redshifts of quasars are a consequence of their great distance.

There is a limit to how large a gravitational redshift can be. The radiation cannot lose more energy than it possesses in climbing the gravitational field of a star. The ultimate redshift occurs when the fraction of energy lost by radiation is 100 per cent, which occurs when the escape velocity at the star's surface is equal to the speed of light. It is possible, then, to conceive of a star with such a powerful gravitational field that radiation cannot leave the star. Nor can matter leave the star since to do so it would have to travel at the velocity of light, and according to Einstein's Theory of Relativity nothing material can travel at the speed of light or faster.

Nothing at all could ever leave such a star because its gravity would be simply too strong—it would be black because no light could leave it, and it would be a hole because anything dropped in could not get out. Hence the name of such a star—the *black hole*.

How to make a black hole

How would such a star be formed? Take an ordinary star (in what is called a "thought experiment," one which is impossible actually to carry out except in imagination) and compress it in a vise in an attempt to make it smaller and thus increase its escape velocity and its gravitational redshift. The star will resist this attempt by increasing its internal pressure—reducing the star's size causes its atoms to pound faster and more often on the jaws of the vise to attempt to pry it apart.

The star will remain in equilibrium in a kind of war between narcissistic gravitational self-attraction and the incest-taboo of its repelling internal pressure forces.

But suppose that the gravitational vise clamped tighter and tighter, squeezing the interior of the star smaller and smaller. A neutron star might be the result. It too would resist further compression but less enthusiastically than ordinary stars: the repelling pressure mechanism is not so easily able to respond to any increased self-attraction. In fact the more massive neutron stars are less able to increase their internal pressure in response to further gravitational contraction. Beyond a certain critical mass in fact, they cannot respond at all: their internal pressure is at a maximum and cannot be increased. The gravitational self-attraction of such a star is always larger than the repelling pressure forces. The star cannot support itself and simply shrinks smaller and smaller in a continuing collapse. As it does so the velocity of escape at its surface increases until it reaches and passes the speed of light, at which stage whatever is inside the surface is shut off from the rest of the Universe forever. The star has become a black hole.

Since, in a sense, a black hole can be formed as a kind of extreme neutron star, it is natural to look to the same process which forms neutron stars to form black holes. Black holes appear to be formed in supernova explosions in the case where the mass of the stellar core which begins the collapse exceeds the critical mass of neutron stars, a few solar masses.

Are black holes the missing mass?

Is there evidence that black holes exist? There is certainly evidence for the existence of invisible matter in our Galaxy. Its effect can be seen in the motions of stars. As it orbits the Galaxy, a star is subject to the gravitational pull of all the others in the Galaxy. The stars which we see away from the Milky Way are above or below the Sun, which lies close to the central plane of the Galaxy wherein lie most stars. Stars above or below the galactic plane are pulled back towards it and the force which pulls them can be estimated from the speeds of stars at different heights above the plane. The higher stars generally move more slowly, just like stones flung from the surface of the Earth. If the force pulling these stars back is known, we can calculate the amount of mass in the galactic plane required to produce such a force.

The answer is that in every cube of space of 10 light years on the side near the galactic plane in the vicinity of the Sun there is on average 4.5 solar masses of material. About 1.8 solar masses of that can be accounted for as visible stars. A further 0.9 solar masses can be detected as interstellar gas in the form of hydrogen atoms. Approximately 2.7 solar masses is invisible. Part of this mass is certainly hydrogen molecules which emit no identifiable radiation received on Earth; part is undoubtedly neutron stars which have stopped being pulsars and part is former white dwarfs which have cooled to invisible black dwarfs. There has been speculation that some fraction at least of this so called "missing mass" in the Galaxy is in the form of black holes. Potentially, then, there may be large numbers of black holes waiting to be found.

But, "of all objects that one can conceive to be traveling through empty space," wrote R. Ruffini and J. A. Wheeler in 1971, "few offer poorer prospects of detection than a solitary black hole . . . " By its very nature the black hole cloaks itself with invisibility. No evidence of it existence beyond its surface can ever be seen, no action on its surface can send a message to proclaim its occurrence. The message carrier—radio pulse, light flash, cosmic ray or what you will—cannot escape the black hole's pull of gravity, which is why the surface of a black hole is called the *event horizon*. Occurrences within it are never seen.

If we cannot see a black hole itself can we see its surroundings? What happens when something encounters a black hole? What would happen if a black hole were to draw material into itself by gravitational force?

How the invisible shines

In an effort to describe how a black hole could be found Iosif Shklovsky in 1967 considered what would happen if a star were to orbit a black hole. The black hole and its companion would circle each other with little effect on each other at first, beyond their mutual orbiting. A distant astronomer might wonder why the radial velocity of the companion star changed periodically and deduce that the star was orbiting another which he could not see. (Such stars are known as single-line spectroscopic binaries and it is possible that some of their invisible companions are black holes, though no doubt the vast majority of invisible companions are simply fainter stars outshone by the star which can be seen.) But there will come a time when the ordinary star circling the black hole will begin to turn into a red giant or supergiant. Its atmosphere will leak onto the black hole. Just as it does when a star's atmosphere falls on to a neutron star, the compressed gas will be heated to temperatures of millions of degrees centigrade and will radiate X-rays. To find a black hole, said Shklovsky, look among the X-ray stars. But how can we distinguish X-ray emissions from a black hole from those from a neutron star?

The major fact which would distinguish the black hole would be its mass, perturbing the companion star by the force of gravity. If the mass of such an X-ray source, measured by the size of the perturbation of the companion star, was larger than the extreme upper limit to the allowable masses of a neutron star, then the X-ray source might be a black hole.

The first such X-ray source known, the invisible companion to a large but ordinary star, is Cygnus x-1. First observed by rocket- and balloon-

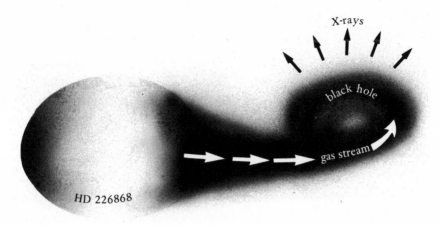

In Cygnus X-1 a large blue supergiant star, HD 226868, may be overflowing onto a companion black hole. Gas from its atmosphere leaks towards the black hole, encircling it before falling in and being heated so as to emit X-rays.

borne X-ray telescopes in the mid 1960s, Cygnus X-1 was one of the first X-ray stars studied by the Uhuru satellite in 1971. With this satellite a group of X-ray astronomers, led by R. Giacconi and H. Gursky, located Cygnus X-1 in a small area of the sky in which radio astronomers (L. Braes and G. Wiley in Holland and R. M. Hjellming and C. Wade in the U.S.) had spotted a radio star which had not been visible before. At the same time that the radio star turned on, Cygnus X-1 changed its X-ray character, proving that the X-ray star and radio star were one and the same object.

At the same position of the radio star was a visible star, numbered 226868 in the Henry Draper catalog (HD). HD 226868 was immediately found to be a hot supergiant star, about 30 solar masses, but not peculiar in any way. The blue supergiant was so normal that two groups of astronomers who studied HD 226868 concluded that it was also a red herring, and nothing to do with Cygnus X-1. But two other groups of

203

astronomers in England and Canada kept observing the star to see whether it changed at all. Tom Bolton at the University of Toronto and Louise Webster and Paul Murdin at the Royal Greenwich Observatory simultaneously published their findings that it did. The star had a cycle of radial velocity change which lasted 5.6 days as it orbited an invisible star, alternately approaching and receding from Earth so that the lines in its spectrum were blue- and red-shifted by the Doppler effect.

It was not possible immediately to say precisely what the mass of the invisible companion was, since the inclination of the orbit of the binary star was unknown and the astronomers could not tell whether they saw the full motion (orbit seen edge-on) or a small part (orbit seen nearly face-on), but clearly the mass of the invisible companion had to be at least six times that of the Sun to swing HD 226868 as it did. This minimum is more than is possible if the invisible companion were a neutron star, being more than the critical mass of neutron stars, namely 3.2 solar masses.

Thus Cygnus x-1 fits in detail the scenario for the discovery of a black hole in a binary star system, outlined by I. Shklovsky in 1967 as a solution before the existence of the problem.

TABLE VII *The velocity of escape*

Earth		11 km/s
Moon		2.4
Mars		5
Jupiter		58
Sun		618
Typical stars:	blue dwarf	900
	red dwarf	400
	blue supergiant	900
	red supergiant	100
Collapsed stars:	white dwarf	3000
	neutron star	150,000

(1 km/s = 2240 miles/hr; speed of light = 300,000 km/s.)

This is not to say that Cygnus x-1 *must* be a black hole. Perhaps the X-ray emitter is a neutron star (of mass 1 solar mass) orbiting a dim normal star (of, say, five solar masses) which itself orbits the bright and visible supergiant. But the black hole explanation accounts for the observed facts and has the attraction that it was proposed before the facts were known; it does not suffer from the suspicion that it has been patched to fit in. Perhaps a black hole, exotic though it may be, has indeed been discovered lurking in Cygnus x-1, formed by a long-past supernova and enabled to shine by devouring its companion.

XV *Final chapter*

Some say the world will end in fire,
Some say in ice,
From what I've tasted of desire
I hold with those who favor fire,
But if it had to perish twice,
I think I know enough of hate
To say that for destruction ice
Is also great
And would suffice.

Robert Frost

The escape velocity is a concept which can be applied to the constituents of the whole Universe as well as to a star or planet within it. All matter in the Universe was subject to the explosion of the Big Bang and may have been given the velocity of escape from the gravitational pull of the rest of the Universe. If this is so, then the energy of motion of the Universe, its kinetic energy, is larger than its gravitational energy and the explosion which began the Universe will never end: the Universe will continue to disperse for ever. If, on the other hand, the Big Bang was not powerful enough to overcome the mutual gravitational attraction of all parts of the Universe, the explosion will eventually coast to a halt and the Universe will collapse; when it gets small enough, it may re-explode and bounce, oscillating indefinitely.

In the technical jargon on the subject, if the gravitational energy of the Universe exceeds the kinetic energy, the Universe is closed and will collapse; if its gravitational energy is less than its kinetic energy, the Universe is open and will expand forever.

206

There are two direct lines of attack on the problem of deciding between these possibilities. The first consists of looking back in time at distant galaxies so far away that they represent the Universe as it was a significantly long time ago, and trying to see what the expansion rate of the Universe was then. The expansion rate may be slowing down so quickly that we can tell whether the Universe will decelerate to a stop and collapse.

To make this method work, astronomers must first measure the cosmic expansion rate of a distant group of galaxies and then determine their distance. It is here that the difficulties arise. If galaxies were all of known intrinsic brightness, like cepheid variable stars, cosmologists could use their apparent brightness as a distance measurement. In the past 20 years, several attempts using this method to determine whether the Universe will expand forever have marginally favored the result that it is closed and will ultimately collapse. But soon after the Big Bang, the galaxies all formed at about the same time, so that distant galaxies are younger than nearby ones (because as we look farther away we look back in time). If younger galaxies are brighter than older ones (because they contain larger numbers of bright stars), they will seem nearer than they really are. Hence we will be measuring the expansion rate for distant galaxies as though they were nearby, and overestimating the amount of deceleration of the Universe.

These results thus seem to be overestimates of the deceleration and may be biased in saying that the Universe is closed.

Another way in which the distance of far galaxies can be obtained is to look at their angular size. This method, applied to clusters of galaxies and to radio galaxies, has given values of the deceleration which are on the borderline between closed and open Universes and are tantalizingly equivocal. They marginally favor the open Universe.

Perhaps a better method of determining whether the Universe is open or closed is to attempt directly to estimate the kinetic and gravitational energy of the Universe to see which is bigger.

In attempting to add up all the mass in the Universe item by item to calculate its gravitational energy, astronomers have come up against the problem that they simply do not know enough about what kind of material predominates in the Universe. Most of the mass of which they are cognizant is in the form of galaxies, and the mass of an average galaxy can be measured in two ways. Astronomers can look at the speed with which stars in the outer parts of a galaxy orbit its center, and estimate the mass required to deflect stars by the amount observed (just as the mass of the companion star to Cygnus x-1 has been estimated by looking at its effect on the visible star HD 226868). Alternatively, they can look at the speeds with which individual galaxies deflect each other when they are situated in a cluster of galaxies.

The former method suffers from the disadvantage that if there is a halo of very faint, undetectable stars surrounding the galaxy, these have no effect on the motion of the visible stars nearer the galaxy's center and they remain undetected. Possibly this is why the two methods for estimating the mass of an average galaxy give answers differing by a factor of 100! Somehow, astronomers may be missing 99 per cent of the matter in an average galaxy!

What form this missing matter takes has been the subject of much speculation. None of the observations exclude the possibility that between or around galaxies lie enormous masses of black holes, faint red dwarf stars, rocks or hot gas (at temperatures of around a million degrees). There is not much cold hydrogen, for this would be seen to play a larger part in absorbing the light from distant quasars than it in fact does. Recent discoveries by the Uhuru X-ray satellite of X-rays from the vicinity of clusters of galaxies has shown that between galaxies in the clusters there does exist very hot gas (at 100 million degrees), but probably not in such abundance that it can close the Universe.

Instead of looking at the components of the Universe item by item and adding them all up to determine the gravitational energy of the Universe, Allan Sandage has attacked the problem by looking at how faithfully the nearby galaxies follow the Hubble law, that their redshifts are proportional to their distances from us. He argues that where galaxies do not follow this law closely, the departures from the law are caused by local clumps of matter (other galaxies, other clusters of galaxies or whatever) and the amount by which they are deflected from the Hubble law tells how much matter is deflecting them. Sandage obtains a result which he tersely summarizes: "Taken at face value these values suggest that (1) the deceleration is almost negligible. . . . (2) the Universe is open, and (3) the expansion will not reverse."

If Sandage is right, the Galaxy will become increasingly isolated from its neighbors as they recede from it. The Galaxy itself will ultimately cease to shine. Already a significant fraction of its mass is locked up in dark stars—white dwarfs and the end products of supernovae: black holes and neutron stars. An increasing proportion of its gas will have processed through stars and increasing amounts of metals will be thrown back into the interstellar medium by supernovae. When the gas gets too metal-polluted, stars which form from it will not be able to shine.

Supernovae not only mark the death of individual stars, they hasten the aging of our Galaxy, possibly towards a dark, cold and lonely death as it finds itself alone in the Universe.

At a conference in Cracow in 1973, John Wheeler conducted an opinion poll of the assembled cosmologists on the question of whether they thought the Universe was closed or open. Of course, truth is not decided by democratic vote—the result of the poll only gives an indication of what most informed people think. Most cosmologists put themselves into the "don't know" category, and were prepared to wait for more solid evidence before pronouncing on the subject. We apparently shall not for a while be able to read the final chapter in the life of our Galaxy, although it is at this moment being written sentence by sentence among the stars, and punctuated by supernovae.

Booklist

Books at a similar level to this one, containing relevant material

N. Calder, *The Violent Universe*. BBC publications. London, 1968.

D. Bergamini, *The Universe*. Time-Life Books. 1964.

F. Hoyle, *Astronomy*. Macdonald. London, 1962.

J. S. Glasby, *Variable Stars*. Constable. London, 1968.

T. Weekes, *High Energy Astrophysics*. Chapman and Hall. London, 1968.

I. Ridpath (ed.), *Illustrated Encyclopedia of Astronomy and Space*. T. Y. Crowell. N.Y., 1976.

Books about supernovae and their remnants available only in specialist libraries, or through the interlibrary loan system

Krishna, M. V. Apparao, *The Crab Nebula*. Astrophysics & Space Science, vol. 25, p. 3, 1973.

P. J. Bancazio and A. G. W. Cameron (ed.), *Supernovae and their remnants*. Gordon and Breach. N.Y., 1969.

D. H. Clark and F. Stephenson, *The Historical Supernovae*. Pergamon. London and N.Y., 1977.

C. B. Cosmovici (ed.), *Supernovae and Supernovae Remnants*. Reidel. Dordrecht, 1973.

R. D. Davies and F. G. Smith (ed.), *The Crab Nebula*. Reidel. Dordrecht, 1971.

Flagstaff Symposium on the Crab Nebula, Publications of the Astronomical Society of the Pacific, vol. 82, p. 375, 1970.

R. N. Manchester and J. H. Taylor, *Pulsars*. W. H. Freeman. San Francisco, 1977.

I. S. Shklovsky, *Supernovae*. Wiley. London, 1968.

I. S. Shklovsky, *Cosmic Radio Waves*. Harvard. Cambridge, 1960.

Index

INDEX